D0504319

How to Make a Human Being

WITHDRAWN FROM STOCK

WITHDRAWN FROM STOCK

WITHDRAWN FROM STOCK

By the same author

*You Are Here: A Portable History of the Universe*

# How to Make a Human Being

A body of evidence

CHRISTOPHER POTTER

FOURTH ESTATE · *London*

First published in Great Britain in 2014 by
Fourth Estate
An imprint of HarperCollins*Publishers*
77–85 Fulham Palace Road,
Hammersmith, London W6 8JB
www.4thestate.co.uk

Copyright © Christopher Potter 2014

1

The right of Christopher Potter to be identified as the author
of this work has been asserted by him in accordance
with the Copyright, Design and Patents Act 1988

A catalogue record for this book is
available from the British Library

ISBN 978-0-00-744779-4

All rights reserved. No part of this publication may be
reproduced, transmitted, or stored in a retrieval system,
in any form or by any means, without permission
in writing from Fourth Estate.

Text design by Richard Marston
Typeset in Minion by G&M Designs Limited,
Raunds, Northamptonshire
Printed and bound in Great Britain by
Clays Ltd, St Ives plc

**MIX**
Paper from
responsible sources
**FSC C007454**

FSC is a non-profit international organisation established to promote
the responsible management of the world's forests. Products carrying the
FSC label are independently certified to assure customers that they come
from forests that are managed to meet the social, economic and
ecological needs of present and future generations,
and other controlled sources.

Find out more about HarperCollins and the environment at
**www.harpercollins.co.uk/green**

For Peter

# Contents

# PART ONE

# Materials and Instructions

What's the matter?

*The opening words of Coriolanus*

# SECTION 1

# Getting started

1 | Clearly one way to make a human being is to start by making a universe of the right kind. But out of what material, and what conditions?

# What can the matter be?

There is only one sort of stuff, namely matter – the physical stuff of physics, chemistry and physiology.
*Daniel Dennett, philosopher*

There is only one kind of stuff in the universe and it is physical. Out of this stuff come minds, beauty, emotions, moral values – in short the full gamut of phenomena that gives richness to human life.
*Julian Baggini, philosopher*

The laws of physics have conspired to make the collisions of atoms produce plants, kangaroos, insects and us.
*Richard Dawkins, biologist*

1 | All of reality is nothing more than an arrangement of particles. Our physical and mental life must be made out of particles because there is nothing else. Everything comes down to what the particles are. Work that out and you know all there is to know.

2 | Bishop Berkeley's* strongest claims to whatever fame is still attached to his name are his theory of immaterialism – that material objects exist only because there is a mind that perceives them – and his

---

* George Berkeley (1685–1753), philosopher, Bishop of Cloyne.

'proof' that there is nothing the world can be made out of. If the world is material and made out of some type of smallest thing, some particle, then whatever that smallest particle is, it must extend into space, since it is in the nature of all material things that they take up room. Furthermore it must be possible in principle, even if we don't know how to do it in practice, to divide these particles into smaller particles; because however small any particle might be, we can imagine some part of it taking up less space. And so the search to find the smallest particles out of which the fabric of the material world is woven must be endless. The argument does not necessarily claim the world as spirit, so much as point out that a material world must be some kind of an illusion: not that the world does not exist, but that it is not what it appears it be.

There is no there there.
*Gertrude Stein (1874–1946), poet and novelist**

I have followed the materialist story of our origin – nay, of my origin. But I have grave misgivings. As an act of faith it requires so much.
*John Eccles (1903–97), neurophysiologist*

I said that the latest advances in science seemed to have made materialism untenable, and that the most likely outcome was still the eternal life of the soul and reunion beyond the grave.
*Marcel – Proust's narrator – to his grandmother in Remembrance of Things Past*

---

* Writing here of her childhood home town of Oakland, California.

# SECTION 3

# Taking sides

1 | Ever since Newton's time, when billiards was in vogue, science has tried to reduce the world to balls hitting one another: billiard ball atoms, billiard ball planets, billiard ball stars. For those of us who have fought shy of games ever since schooldays, it is sometimes hard to accept that ball games really are the be all and end all of existence. Even on those days when I know – or is it fear? – that all that there is can be reduced to particles, I am dispirited. I feel as I did at school: I know that materialism is the manlier choice, but it just isn't me.

There are days when the world seems to be split into two teams and I do not know which side to be on. Brian Greene, Richard Dawkins (captain), Daniel Dennett, Dr Johnson, Thomas Jefferson, Lucretius, Stephen Hawking, Aristotle, David Attenborough and Thomas Huxley are on one side.* Marcel Proust, Leo Tolstoy, William James (captain), Marilynne Robinson, John Keats, Rowan Williams, Karen Armstrong, Plato, William Blake and Emily Dickinson on the other. Dr Johnson and Rowan Williams sometimes play in goal. Proust and Keats invariably call in sick. Darwin and Descartes have been known to show up for either side. Einstein is a popular referee. Confusingly, there are times when it is hard to tell which team even Richard Dawkins or Brian Greene is playing for. But generally Richard Dawkins's team terrifies me and wins. William James's team invariably loses, but they don't seem to care.

---

* No women! See page 9, 5 ii

On the one side are the materialists: what you see is what you get. The world can be reduced to basic ingredients, and those ingredients are material: they exist, can be weighed and counted, measured and timed. Materialism, reductionism: words that shine with confidence.

On the other side are the idealists, who believe that the physical world is somehow a manifestation of something immaterial. We are the transcendentalists, they cry ('Give us a T ...'). Idealism, transcendentalism: words that sound airy-fairy.

2 | At school I remember games period, lining up, waiting to be chosen, down to the last four, the final humiliation of being the very last hardly averted by the gamesmaster: 'The rest of you just divide up equally'; then the desperate rush to attach myself to what I hoped was to be the stronger side, wanting to be on the winning team but not wanting to take part, hoping that I might be in goal, left alone to sing hymns to myself while everyone else battled it out at the other end of the pitch.

3 | I remember, too, poring over a copy of the Ladybird book of Roundheads and Cavaliers, puzzled. Clearly, my heart told me, it was better to be a Cavalier: the clothes, the hair, the colours! And yet rationally I knew that to be a Roundhead was the right, the moral choice.*

4 | There are these days:

> I am satisfied and sufficiently occupied with the things which are, without tormenting or troubling myself about those which may indeed be, but of which I have no evidence.
> *Thomas Jefferson (1743–1826), American founding father*

---

\* Years later when I read *1066 and All That* by Sellar and Yeatman I discovered that I had not been alone in my dilemma: 'Wrong but romantic' they characterised the Cavaliers, 'right but repulsive' the Roundheads.

There was no room for the mystical, the mysterious, the illusory in his temperament. Anything that did not stand the test of practical experience or the scrutiny of analysis he rejected as an optical illusion, some kind of interplay of light and colour on his retina, or else a phenomenon that still lay beyond the reach of experience. There was in him nothing of the dilettante who loves to delve into the realm of the fanciful and idle speculation about the wonders and marvels that lie a thousand years into the future. He took a firm stand on this side of the threshold of the mysterious, free equally of a childlike credulity and the doubts of the over-sophisticated, and patiently reserved judgment until the evidence came in and provided a key to the mystery.

*Konstantin Levin in Tolstoy's Anna Karenina*

5 | There are also these days:

i Days when I feel like Fotherington-Thomas* – 'who sa hello clouds hello sky', and who 'like all goody-goodies he believe in fairies father xmas peter pan etc and unlike most boys they are kind to their sisters' – days when 'I simply don't care a row of buttons whether it was a goal or not nature alone is beautiful.'

ii Or like the evolutionary biologist Richard Lewontin when he writes that science is 'filled with the violence, voyeurism, and tumescence of male adolescent fantasy. Scientists "wrestle" with an always female nature, to "wrest from her the truth", or "to reveal her secrets". They make "war" on diseases and "conquer" them. Good science is "hard" science; bad science (like the refuge of so many women, psychology) is "soft" science, and molecular biology, like physics, is characterised by "hard inference". The method of science is largely reductionist, taking Descartes's clock metaphor as a basis for tearing the complex world into small bits and pieces to understand it, much as the

---

* In *How to be Topp* (1954) by Geoffrey Willans and Ronald Searle.

archetypal small boy takes apart the real clock to see what makes it tick.'

6 | But there are days, perhaps most days, when it is not clear which side to be on.

7 | When told of Bishop Berkeley's fashionable new philosophy of immaterialism Dr Johnson* said, 'I refute him thus,' and kicked a rock. I've always had a soft spot for Dr Johnson. He preferred people to places, and loved his cats. When he first read *Hamlet* he was so frightened by the ghost of Hamlet's father that he rushed outside in order to have living people about him. According to his friend Jonas Hanway, he was 'a hardened and shameless tea-drinker'. He devoted himself to conversation. He was terrified of eternal damnation. As an entertainment to his guests, he pulled up his tailcoats to form a pouch and jumped about in imitation of a kangaroo. When he was four years old his mother called him a puppy and he said to her: 'Do you know what they call a puppy's mother?' He rolled down a hill in Lincolnshire when he was in his fifties. He leapt a wall in his seventies. When William Hogarth met him he mistook him for an idiot, 'shaking, twitching, pock-marked, half-blind and distinctly careless about his dress'. He compiled a dictionary of the English language from scratch, and although after two years he got stuck on the word 'carry', he persevered. In 1755, nine years after he had begun, the dictionary that made his name was published in two volumes, each volume weighing in at fourteen pounds. He said he had tried studying philosophy, 'but cheerfulness was always breaking in'. In that unsubtle, if entertaining, gesture Johnson kicks more than a rock; he directs a kick in the general direction of all philosophical argument that goes against common sense. And it's common sense, the gesture insists, that makes us human, and not some illusive philosophical argument. Dr Johnson kicked His Grace into the sidelines of history.

---

* Samuel Johnson (1709–84), man of letters.

8 | And yet, if Dr Johnson has about him the no-nonsense mien of the Roundhead, it is surely as a shield to protect the heart of a Cavalier. Tolstoy's Levin could hardly be less like Dr Johnson (for one thing he is fictional), but he too combines traits of both Cavalier and Roundhead, idealist and materialist. Levin wants to believe but cannot. Yet it is his unbelief that tortures him. As a child and adolescent he had turned to Christianity to try to address the questions of life: 'whence it came, wherefore, why, and what it was'. Finding no answers there, he turned as a young man to science: 'involuntarily, unconsciously, he now sought in every book, in every conversation, in every person, a connection with these questions and their resolutions'. By his mid-thirties he found that even scientific answers no longer satisfied him: 'he became convinced that those who shared the same views with him simply dismissed the questions which he felt he could not live without answering ... He was in painful discord with himself and strained all the forces of his soul to get out of it ... He read and pondered, and the more he read and pondered, the further he felt himself from the goal he was pursuing ... Convinced that he would not find an answer in the materialists, he reread, or read for the first time, Plato, and Spinoza, Kant, Schelling, Hegel and Schopenhauer – the philosophers who gave a non-materialistic explanation of life'. But nor does philosophy bring consolation: 'Following the given definition of vague words such as *spirit, will, freedom, substance* ... he seemed to understand something. But he had only to forget the artificial train of thought and refer back to life itself ... and suddenly the whole edifice would collapse like a house of cards.'

> But Levin did not shoot himself or hang himself and went on living ... not knowing and not seeing any possibility of knowing what he was and why he was living in the world, tormented by this ignorance ... and at the same time firmly laying down his own particular, definite path in life.
>
> *Leo Tolstoy (1828–1910), Anna Karenina*

# Nothing → something → everything

1 | Physicists don't pack up just because some philosopher or other points out that what they are doing is paradoxical. Particle physicists accept that elementary particles are not required to extend into space, nor to exist. If that means redefining what it means to exist, then so be it.

2 | In the main, scientists working at the coalface have little time for the fine distinctions made by philosophers. They are interested in what is in front of them. They are Dr Johnson not Bishop Berkeley, Romeo not Friar Laurence.

3 | Philosophy is adversity's sweet milk, says Friar Laurence. Hang up philosophy! says Romeo in reply. Unless philosophy can make a Juliet, Displant a town, reverse a prince's doom, It helps not, it prevails not: talk no more.

> Philosophy is dead.
> *Stephen Hawking, on the first page of his book The Grand Design*

> Philosophy is to science what pornography is to sex.
> *Steve Jones, biologist*

Philosophers keep out. Work in progress.
*A notice pinned to the laboratory door of the physicist Niels Bohr (1885–1962)*

If you ask in how many cases in the past has a philosopher successfully solved a problem, as far as we can say there are no cases.
*Francis Crick (1916–2004), biologist*

4 | In refutation of Zeno's paradox,* Diogenes got up and walked across the room.

To study Metaphysics as they have always been studied appears to me to be like puzzling at astronomy without mechanics.
*Charles Darwin (1809–82), in his notebook*

Sometimes he thought sadly to himself, 'Why?' and sometimes he thought, 'Wherefore?' and sometimes he thought, 'Inasmuch as which?' – and sometimes he didn't quite know what he was thinking about.
*Eeyore the philosopher in A.A. Milne's Winnie the Pooh*

---

* In a series of paradoxes, the ancient Greek philosopher Zeno described various impossibilities of motion. For example, most famously: in a race, a tortoise is given a head start over a hare. At some point in time the hare will reach the place where the tortoise began, but by that time the tortoise will have moved some distance ahead. And by the time the hare reaches that further point, the tortoise will again have moved on. The tortoise's lead has been severely reduced, but no matter at what time the hare gets to where the tortoise has been, the tortoise will be some small, if reducingly small, distance ahead. It would appear that it is impossible that the hare will ever overtake the tortoise. The resolution of the paradox is in the understanding that an infinite sum of reducing elements is not itself always infinite. Many infinite series are finite when added together. Any student of mathematics knows that the sum that stretches to infinity $1+\frac{1}{2}+\frac{1}{4}+\frac{1}{8} \ldots$ is not infinity as the ancient Greeks supposed, but 2. How to integrate together infinite sequences that contain infinitely small elements is at the heart of calculus. In *War and Peace* Tolstoy devotes a chapter to calculus in order to illustrate poetically how the motion of history is the integral of the anonymous individuated lives of the world's population, not the force of will of a single man, even if that man is Napoleon.

Philosophers have been profoundly wrong in almost every question under the sun over the last 2,000 years. You should never listen to the answers of philosophers, but you should listen to their questions.
*Christof Koch, neuroscientist*

Philosophers will tell you the whole idea of science is just a subset of philosophy.
*David Rothenberg, philosopher*

5 | Philosophy used to matter more. Of Plato's five kinds of imagined regimes, the greatest – named Kallipolis – was ruled by philosopher kings.* Socrates had to die because philosophy was seen as a threat to society. These days philosophy matters only to philosophers.

Philosophy is the highest, the worthiest, of human endeavours.
*Slavoj Žižek, philosopher*

6 | Scientists have a habit of dismissing the questions they don't want to answer. They call them philosophical. For many scientists, philosophy is a step too close to theology. Scientists eschew philosophy for logic or even for just plain common sense.

The whole of science is nothing more than a refinement of everyday thinking.
*Albert Einstein (1879–1955)*

It is generally thought that common sense is practical. It is practical only in a short-term view. Common sense declares that it is foolish to bite the hand that feeds you. But it is foolish only up to the moment when you realize that you might be fed very much better.
*John Berger, writer and art critic*

---

* Tyranny was the worst of the regimes, and democracy the next to worst.

7 | If, as philosophers have concluded, there is nothing that the universe can be made out of, scientists have wondered what that nothing might be.

8 | Science, philosophy and religion have this in common: that they all must account for how nothing became something. Philosophers have worried for centuries about the nature of substance and of nothingness. Religions have their various creation myths. Science too tells its own creation story. The triumph of particle physics is that it so nearly explains how nothing became everything.

> Why is there something rather than nothing?
> *Gottfried Leibniz (1646–1716), philosopher and mathematician*

> All things are born of nothing and are borne onwards to infinity.
> *Blaise Pascal (1623–62), philosopher and mathematician*

9 | If only we knew exactly how the story began. If only we could say, 'Once upon a time' and know what follows.

10 | An ancient Greek cosmology has the world created out of a pre-existing condition called chaos. *Kaos* is not emptiness but formlessness. It was the world before there were things in it. The word nothing, like some fossil of ancient thought, still retains that original concept of no thing. The universe emerged when *Logos*, meaning variously form, knowledge and word, came into contact with *Kaos*. Out of their union comes *Cosmos* (beauty or order, as in cosmetics, which bring order to the face). The opening words of St John's gospel repeat this ancient prescription: In the beginning was the Word. In the original Greek the word translated as 'word' is *Logos*. And in Genesis we find God creating nature by separating out from chaos what then become *things* with names. Naming is a process of separating out, and the first step in any scientific investigation of the world. Before explanation must come the naming of parts. The idea of nothing as emptiness came later. That the

universe was created out of emptiness, *ex nihilo*, is a radical departure from how creation was envisaged by certain ancient Greek philosophers, and was an interpretation imposed on the Biblical story by medieval scholars.

11 | Our current best modern-day creation stories are variants of the Big Bang theory, a mathematical description of the universe coaxed out of the equations of general relativity. Even though they were his equations, Einstein at first denied the Big Bang. Later he changed his mind.

> The most beautiful and satisfactory explanation of creation to which I have ever listened.
> *Einstein, of an exposition of the Big Bang given by its inventor (discoverer?) Georges Lemaître (1894–1966), priest and physicist*

12 | All matter can ultimately be reduced to constituent particles – bosons and fermions – that are the excitations of various types of energetic field. At the Big Bang there may have been a single kind of energetic field which, in an expanding universe, evolved into other kinds of energetic field.

Within a trillionth of a trillionth of a trillionth of a second after the Big Bang the universe is a cascade of particles decaying into other particles. Whole eras of the universe passed before it was even a second old.

13 | We know what happened in the first trillionth of a trillionth of a trillionth of a second, but what happened in the beginning?

14 | In the beginning everything was in the same place at the same time. In the beginning the physical world is pure energy, whatever that is. In the beginning the universe is some condition of form, number and energy held in perfect symmetry. It cannot last. The symmetry breaks and becomes a world of asymmetries, imperfections and accidents. The world falls into existence.

The positive energy within matter can be counterbalanced by the negative sink of the all-pervading gravitational field such that the total energy of the universe is potentially nothing; when combined with quantum* uncertainty,† this allows the possibility that everything is … some quantum fluctuation living on borrowed time. Everything may thus be a quantum fluctuation of nothing.

*Frank Close, particle physicist*

Zero exists now, it has always existed, and it will always exist. It is the native state of existence. It is what the physicist David Bohm called implicate order. It is the timeless quantum superposition of all universes and all life in an infinite universe. As the most brilliant physicists have long held, a perfect zero is the most ordered state of all, it just isn't found in the past where time begins. It exists in the future where time ends.

*Gevin Giorbran, science writer*

15 | Energy leaks out of the vacuum for no reason at all except randomness and the pressure exerted by a sink of infinite negative energy. Overall the universe is nothing at all.

16 | Take matter out of the universe and reality becomes unstable, liable to give birth randomly to new universes. The vacuum is the birthing ground of universes; like the silence of the mystics, a roiling place of visions and madness, of annihilating forces.

---

\* Quantum mechanics tells us that reality is woven out of many packages – called quanta – of some smallest possible amount of energy.

† In 1927 the German physicist Werner Heisenberg propounded his famous 'Uncertainty Principle'. It shows that a complete classical description of reality as Newton envisioned it is impossible. Reality can be known only uncertainly, not precisely.

17 | In the outer reaches of the universe, as far away from here as it is possible to be, beyond time and space and meaning and matter, nothing was happening. And the nothing was without form, pure potential for becoming, an evanescent yet heaving sea of energy coming into and out of existence. For reasons not yet understood, a bubble of energy that should have burst back into non-existence breaks free with the rage of Achilles from the conditions of the quantum world and sweeps out a universe.

18 | The universe is just one of those things that happens from time to time. Everything that is exists only by happenstance, randomly, out of nothing.

19 | If less is more, is nothing too much?

20 | For now the most widely-agreed-on model that describes how the universe got going is the theory of eternal inflation. An infinite number of 'bubbles' arose in an eternally inflating quantum landscape. One of these bubbles became the island universe we call home. An infinite number of other island universes exist in all the possible forms determined by some constraining mathematical model, most popularly string theory.* The landscape out of which these island universes emerged is called the multiverse.

21 | Inflation is happening eternally, elsewhere. Our 'island' universe inflated briefly. It doubled in size every $10^{-34}$ seconds. After about a hundred such doublings it had grown to about the size of a grapefruit, at which point the period of inflation came to an end.† Why inflation came to an end locally is not yet known.

---

* String theory is not very constraining. It presents an array of some $10^{500}$ different initial conditions.

† The universe is, nevertheless, infinite, regardless of whether the so-called visible universe is the size of a grapefruit or billions of light years across.

I could be bounded in a nutshell, and count myself king of infinite
space.
*Shakespeare (1564–1616), Hamlet*

22 | i You do not need God in order to create a universe, says Stephen
Hawking. All you need is gravity, quantum electrodynamics, special
and general relativity, M-theory and a few other bits and pieces of physics. But where these ingredients come from remains, for the moment at
least, an unanswered question.

ii M-theory (no one can remember what the M stands for) is a formulation of string theory, and a quantum theory of gravity – an abstract
and mathematical theory, as yet without physical proof. M-theory
describes a multiverse of eleven dimensions in which there may be
many island universes like ours, adrift in four dimensions of time and
space; and many other kinds of universes adrift in different numbers
of dimensions of space, some perhaps with several dimensions of time
(whatever that might look like). M-theory describes $10^{500}$ different
universes.

Consider the most obvious question of all about the initial state of
the universe. Why is there an initial state at all?
*Lawrence Sklar, philosopher of physics*

The desire to find a beginning comes from the idea that everything
has the real, solid existence that our minds generally perceive.
*Matthieu Ricard and Trinh Xuan Thuan, The Quantum and the
Lotus*

As far as I can see, such a theory [as the Big Bang] remains entirely outside any metaphysical or religious question. It leaves the materialist free to deny any transcendental Being … For the believer, it removes any attempt at familiarity with God. It is consonant with Isaiah speaking of the hidden God, hidden even in the beginning of the Universe.

*Georges Lemaître*

23 | In saying that the universe randomly evolved out of some initial energy condition that we don't yet fully understand, we sweep everything we don't know about the universe under the carpet. All the unanswered questions about the physical universe get pushed to its horizons, far away from where humans are. The horizons of the universe are the limits of what we can see and what we can understand. The universe disappears over its own horizon, taking with it the laws of nature, forever just out of our reach. For a while, the more we found out about the physical universe the larger it became. But largeness itself has become *passé*. The universe shows itself to be subtler than mere size. All our creative speculations, even when they harden into theories, merely push the mystery of what we are and where we come from to ever more distant regions of an ever more elusive universe.

# SECTION 5

# What is science?

Science (a term in itself inoffensive and of indefinite meaning).
*Joseph Conrad (1857–1924), The Secret Agent*

That bright-eyed superstition known as infinite human progression.
*Terry Eagleton, literary theorist and critic*

1 | Life may be messy, but in the physical world there appears to be underlying order. Evidence of this order has encouraged scientists to believe in the existence of physical laws of nature. Why nature should have unifying features is a deep mystery. That physical laws of nature are ultimately reducible to mathematics is an even deeper mystery.*

2 | The difference between the ways of science and the ways of other truth-seeking enterprises is that science has a method.

First find what you think might be a solution to a problem, then express it as a mathematical model, then test it.
*David Deutsch, physicist*

---

\* And there is a further mystery. If the physical world can be reduced to mathematical equations, 'What is it,' in Stephen Hawking's memorable phrase, 'that breathes fire into the equations and makes a universe for them to describe?'

3 | In science, to look is not enough, there needs also to be intervention in order to affirm what it is that is being looked at. A testable theory is required, not just mere description, though a description is a start. A theory is proven for as long as it is confirmed in that repeatable process of measurement called experiment. Sometimes we improve our ability to measure and theories are further confirmed, and sometimes theories fail when examined more closely.

If the explanation of physical phenomena were evident in their appearance, empiricism would be true and there would be no need for science as we know it.
*David Deutsch*

'We admit the existence of electricity, which we know nothing about, why can't there be a new force, still unknown which ...'

'When electricity was found,' Levin quietly interrupted, 'it was merely the discovery of a phenomenon, and it was not known where it came from or what it could do, and centuries passed before people thought of using it. The spiritualists on the contrary, began by saying that tables write to them, and spirits come to them, and only afterwards started saying it was an unknown force.'

Vronsky listened attentively to Levin, as he always listened, evidently interested in his words.

'Yes, but the spiritualists say: now we don't know what this force is, but the force exists, and these are the conditions under which it acts. Let the scientist find out what constitutes this force. No, I don't see why it can't be a force, if it ...'

'Because,' Levin interrupted again, 'with electricity, each time you rub resin against wool, a certain phenomenon manifests itself, while here it's not the same each time, and therefore it's not a natural phenomenon.'

...

'I think,' he continued, 'that this attempt by spiritualists to explain their wonder by some new force is a most unfortunate one. They speak directly about spiritual force and want to subject it to material experiment.'

They were all waiting for him to finish, and he felt it.

'And I think that you'd make an excellent medium,' said Countess Nordston, 'there's something ecstatic in you.'

*Leo Tolstoy, Anna Karenina*

4 | The astronomer Nicolaus Copernicus (1473–1543) wondered if there might be some sort of inclination in matter that causes matter to be drawn to itself. But his was a vague poetic notion; it didn't have what elevates Newton's description – of what was later named gravity – to the level of theory. Newton writes in mathematics how the force works even as he fails to tell us what it is. Since it acts at a distance without any visible means of action, Newton's gravity has no material existence, for which lack the theory was criticised by followers of Descartes, who believed that physical actions must result only from physical causes. But in physics mathematics trumps material means. The theory works, and that is enough, particularly when the theory is as encompassing as this one: explaining as it does both why an apple falls to earth and why the earth falls perpetually towards the sun.

A force without materiality looks indistinguishable from magic, but mathematics is what makes the reality of gravity testable. The lack of visible means was first criticised, then overlooked, and finally forgotten about.

If they don't depend on true evidence, scientists are no better than gossips.

*Penelope Fitzgerald (1916–2000), The Gate of Angels*

5 | Science is an attempt to make knowledge collective. Science separates out from the world what can be repeated. Scientific experiments are repeatable (in theory at least) by anyone, ideally not just any

competent human but any competent alien. Art is collective evidence of shared experience too, but science goes further; its knowledge means to be universal, not 'merely' human.

6 | Science searches for evidence of stability in the world out there. At one time we saw stability in the so-called fixed stars, until it was discovered that they are not fixed, just moving very slowly relative to each other, and only appear fixed because they are so far away. We used to think space and time were immutable, until Einstein showed otherwise. Today we begin to wonder if even the speed of light is a constant.

7 | Science organises the meaning of the world into what it is hoped are irreducible statements called the laws of nature, but every seemingly irreducible statement is doomed ultimately to be replaced by another attempt at the irreducible; and so science makes progress. There are no truly fundamental theories in science. Something more fundamental always comes along, eventually. 'Fundamental' theories are the theories that are currently most effective, but they are never complete. And never being truly fundamental, we cannot know if they are ever truly universal. In science, to understand more deeply is to get under what it is that is currently being stood on. What science stands on is continually being replaced by lower floors.

8 | Science has this particular strength, that theories are only overthrown when a new theory encompasses more phenomena than the previous theory encompassed. In this sense, old theories do not die, a new theory reveals the limits within which the old theory was, and still is, effective; but crucially the new theory goes beyond those boundaries into territory in which the old theory fails. Some scientists say that Newtonian physics was shown to be incorrect by Einstein's theories of relativity, others that Einstein showed the limits within which Newton is true.

9 | Truth in science is a comparative entity: there is always the possibility of truer, but what is truer may sometimes – not always, by any means – look completely different from what was almost as true. In order to encompass all that has gone before, a new theory may – occasionally – be forced to do so by completely refashioning the nature of reality.

10 | Einstein twice had to rearrange the material world in order to satisfy the demands of mathematics. In order to take James Clerk Maxwell's equations seriously – they describe electric and magnetic fields and their relationship to each other – he showed that it was necessary to change our thinking about what motion is. The invariant nature of light, he asserted, sets a limit on how fast things can move. In our everyday world we assume that however fast we travel, someone or something might conceivably travel faster; but this is not how things are, only how they appear to be at the relatively slow speeds (compared to the speed of light) of everyday human life. The classical world of Newtonian mechanics is limited, and Einstein shows where and why. No doubt Einstein's assumption about light will prove to be an approximation too, but we do not yet understand how. When we do we will call it progress.

Einstein's assumption upheaved time and space into a conjunction that Nabokov called 'that hideous hybrid whose very hyphen looks phoney'. 'Space-time' more accurately describes the nature of reality than space and time taken separately. And yet, puzzlingly, we do not experience the space-time continuum. We believe we experience space and time separately. But since we do not have sense organs devoted to experiencing either space or time, perhaps what we experience comes from habit or is a delusion. In any case, all our best measurements show that space-time is a better approximation to reality than Newton's theatre set in space and time.

In 1900, Max Planck solved a seemingly intractable problem in physics by breaking light into small packets. He effected a revolution in science even though he did not personally believe that these small packets – later named quanta – were anything more than a mathematical trick. Einstein took quanta seriously, and effectively invented quantum physics (aka quantum mechanics).

11 | The magnetic moment of an electron has been measured to eleven decimal places, and is still in agreement with what quantum mechanics predicts it should be. There is no guarantee that in the twelfth decimal place some new theory might not be required.

> Astronomers have been monitoring the orbits of one double neutron star system – known as PSR 1916+16 – for around forty years. The emission of Einstein's predicted gravitational waves from this system has been confirmed through a very gradual shortening of the star's orbital period, and there has been agreement between the signals received from space and the overall prediction of Einstein's theory to an astonishing fourteen decimal places.
> *Roger Penrose, mathematical physicist and philosopher*

In the fifteenth decimal place, who knows?

12 | Finer measurement requires the invention of finer measuring instruments. Finer measurements lead to better theories. Out of better theories come subtler experiments. Progress is what results from the positive feedback of theory, experiment, and tool-making.

> Very little in nature is detectable by unaided human senses. Most of what happens is too fast or too slow, too big or too small, or too remote, or hidden behind opaque barriers, or operates on principles too different from anything that influenced our evolution. But in some cases we can arrange for such phenomena to become perceptible, via scientific instruments.
> *David Deutsch*

13 | Science goes in search of what is truer. If that means science goes in search of the truth, it does not logically follow that the truth exists, nor, if it does exist, that the truth can ever be reached. Even if we did believe that some ultimate truth exists, we can have no idea what it might look like.

14 | When Copernicus first wondered if the sun, and not the earth, was at the centre of things, he stumbled on what was to become a powerful driving force of the scientific method: that human beings do not occupy any position of privilege in the universe. Scientific progress is the attempt – repeated over and over again – to remove human beings from the centre of things. We are not at the centre of anything, is what the scientific method continually reminds us.

15 | The centre is a place of privileged perspective. If science is to find universal laws, then by their very nature universal laws cannot be privileged, or they would not be universal; and if they are not universal – no matter how grand they are – they are provincial.

16 | In order to uphold its own central tenet that humans are not at the centre of anything, the scientific method itself must be universal. It must have been discovered elsewhere by other intelligences across the universe. Science needs aliens. The more aliens the better. That no aliens have so far stepped forward might be seen as a blow to materialism.

> We can be sure that any intelligent beings inhabiting those planets
> will measure the same inverse square law.
> *John Gribbin, science writer*

It will be a great day in the history of science if we sometime discover a damp shadow elsewhere in the universe where a fungus has sprouted. And here we are, a gaudy efflorescence of consciousness, staggeringly improbable in light of everything we know of the reality that contains us.

Marilynne Robinson, *Absence of Mind*

17 | If aliens do not exist, the whole question of what kind of reality science describes is called into question. So long as aliens do not exist, what we are as humans in the universe remains an open question. So long as aliens do not exist, human beings are the aliens in the scientific woodpile.*

18 | There is nothing inevitable about the way the scientific method has developed on earth. Nor is it inevitable that human beings should have stumbled on the scientific method at all. Many civilisations have come and gone with very different cosmologies. And if not inevitable here, there may be many other worlds where there is sentient life but no scientific understanding. Presumably there are cosmologies out there, as there have been here on earth, that come at the universe from quite a different perspective. Why should other life forms care about intelligence most of all? They may have discovered other motors of the universe. Scientific progress is directly related to our ability to imagine what alien life might be.†

---

* Ironically, there is vast anecdotal evidence for aliens, but as yet no scientific evidence. John Mack, Professor of Psychiatry at Harvard, has said that taking aliens seriously does not mean the anecdotal evidence needs to be considered literally, but since it must be evidence of something, and there being so much of it, it should be taken seriously. By opening up our consciousnesses we may, he says, see the cosmos 'filled with beings, creatures, spirits, gods … that have through the millennia been intimately involved with human existence'.

† A vinyl LP recording of Bach's Second Brandenburg Concerto was sent on two separate Voyager space programmes as a gift to potential alien interceptors. Just a few decades later, the gesture already looks parochial.

Our conceptual model of space and time has proven to be extremely successful – to such an extent that we may even find it difficult to imagine other ways of organising our thoughts and experience – but it isn't logically inevitable.
*Marilynne Robinson, novelist and essayist*

[Theories] about the nature of the world become frameworks within which we live. And so they constrain what we think is possible, what we think is real.
*Max Velmans, professor of psychology*

19 | Scientific theories are frameworks that attempt to contain the world. But these frameworks are never more than detailed models, and a model is not the thing itself. Any knowledge about something is not the thing itself either. The only complete model of the universe would be the universe.

There is nothing more deceptive than an obvious fact.
*Sir Arthur Conan Doyle (1859–1930), 'The Boscombe Valley Mystery'*

20 | Scientific facts are attached to theories that are part of a methodology that continually changes the limits within which they can be regarded as true. Facts are always embedded in theory, and theories come and go. There are no facts without theories. What we take to be facts may not be facts for other intelligent, questioning life forms. We can know the facts, but why they mean anything is another matter.

21 | It is far from clear that there are universal laws,* but the pursuit of them has resulted in what we call progress, the outward and visible evidence of which is the material world we live in.

---

* The theoretical physicist John Wheeler (1911–2008) once questioned the belief in immutable laws, and speculated that in the future we might discover that the laws

22 | The scientific fundamentalist takes home with him his belief that laws of nature actually exist. The more open-minded scientist understands that science is a kind of game whose rules only need apply in the laboratory or at the desk.

23 | Science is a mode of enquiry, not the last word. There are different world views, and they do not have to be commensurate, or agree with each other. And you don't have to say one is better than another in all domains. But clearly if you want to build a rocket you will turn to physics, not theology.*

> And why, after all, may not the world be so complex as to consist of many interpenetrating spheres of reality, which we can then approach ... by using different conceptions and assuming different attitudes.
> William James (1842–1910), *The Varieties of Religious Experience*

24 | There are new theories to come that are beyond the reach of current technologies and of our current imaginations. In order to make progress sometimes a technological leap will come first, as it did when the telescope turned from plaything into scientific measuring instrument. Sometimes experiment comes last of all, as it did when Einstein re-imagined gravity as the geometry of space-time. He spent ten years working out the mathematics, leaving it to others to prove by experiment that gravity was indeed how he conceived it to be. When Einstein was asked what his response would be if experiment were to prove his theory false, he said he would feel sorry for the dear Lord.

---

themselves are flexible and evolving, a conjecture that in recent times has been elaborated by the cosmologist Lee Smolin.

* And turn first to physics and biology if you want to make a human being.

25 | Experiments are generally hard to perform and require determination. No one would perform an experiment without already having some idea of what they are looking for.

> Every brilliant experiment, like every great work of art, begins with an act of the imagination.
> *Jonah Lehrer, writer*

26 | For there to be progress in science there has to be some kind of understanding that comes in advance of the finding out: intuition. The history of science is necessarily full of instances in which insight comes first, ahead of proof in observation and theory. Where does, where can that insight come from? There must be various conduits of the truth if imagination sometimes gets there first.

In science the leap of imagination must be of the right kind and not too great a leap. Mediums and other sensitives also claim the ability to see ahead of the material evidence, but their methods fail when exposed to scientific, repetitive investigation. Their evidence is personal and anecdotal, not public and repeatable as science demands.

27 | On the radio I hear the announcer describe the discovery of a new planetary system as 'a rather wonderful poetic idea'. And why not?

28 | In a purely material world the immaterial is what we don't yet understand materially; a dwindling pile in the to-do basket of science. If we wait too long the ink will have faded and the mystery will have become illegible. How long we are prepared to wait for material answers to material questions tests our faith.

We call the boat back in – Come in, number 87, your time's up – only to find that the boat is too far out, and anyway, if we but knew it, the boat long ago rotted away and sank without trace.

29 | We don't know what Nature is. There is that sifted-out part of Nature we call the material world, that ongoing conversation between science and the world, and then there is the world itself, in the largest sense, in which we are embedded. Most of us, most of the time, confuse the material world with the real world, whatever that is.

The scientific method sieves out the material world. The question is left open whether or not there will be anything left in the sieve afterwards, or indeed, if there is an afterwards. It seems increasingly likely that science may at best describe its own limitations, and not 'everything', as is sometimes predicted by its fundamentalists.

> 'Freddy,* I'm told that there are left-overs in the larder. Have you any idea what to do with left-overs?'
>
> 'You don't have to do anything with them. They're left over from whatever was done to them before.'
>
> His father smiled and sighed.
>
> *Penelope Fitzgerald, The Gate of Angels*

30 | Materialism describes a world made out of logic and things that move. If it cannot be measured by a clock and a ruler, it lies outside scientific enquiry. That the whole world is capable of being measured requires faith, and there are days when my faith falters.

---

* Freddy is a physicist, his father a vicar.

SECTION 6

# What is the universe?

What is the universe but the question what is the universe?
*A line seen at an exhibition at the American Museum of Natural History*

That tempting range of relevancies called the universe.
*George Eliot (1819–80), Middlemarch*

The universe we observe has precisely the properties we should expect if there is, at bottom, no design, no purpose, no evil and no good, nothing but blind pitiless indifference.
*Richard Dawkins*

Nothing: a mechanical chaos of casual, brute enmity, on which we stupidly impose our hopes and fears.
*John Gardner (1933–82), Grendel*

In the Inflationary Multiverse, our universe could well be an island oasis in a gigantic but largely inhospitable cosmic archipelago.
*Brian Greene, theoretical physicist*

It doesn't seem to me that this fantastically marvellous universe, this tremendous range of time and space and different kinds of animals, and all the different planets, and all these atoms with all their motions, and so on, all this complicated thing can merely be a stage so that God can watch human beings struggle for good and evil – which is the view that religion has. The stage is too big for the drama.

*Richard Feynman (1918–88), physicist*

Our conception of the significance of humankind in and for the universe has shrunk to the point that the very idea we ever imagined we might be significant on this scale now seems preposterous.

*Marilynne Robinson, Absence of Mind*

The universe is not there to overwhelm us; it is our home, and our resource. The bigger the better.

*David Deutsch*

I don't think it makes sense to give more importance to a mountain than an ant.

*Joan Miró (1893–1983), painter*

The stars are crushing, but mankind in the mass is even above the stars.

*W.N.P. Barbellion (1889–1919), diarist*

Once I was beset by anxiety … I could have cried out with terror at being lost. But I pushed the fear away – by studying the sky … I saw myself in relationship to the stars. I began weeping, and I knew that I was all right. That is the way I make use of geometry today. The miracle is that I am able to do it with geometry.

*Louise Bourgeois (1911–2010), artist*

Now I see that great men have no other function in life except to help us see beyond appearances: to relieve us of some of the burden of matter – to 'unburden' ourselves, as the Hindus would say.

*Jean Renoir (1894–1979), film director*

1 | How do we unburden ourselves of the weight of the material world pressing down on us? By confession, meditation, losing the ego, giving stuff away, selflessness? I wonder if the apparent joy of the weightlessness of being in outer space is in part the feeling of being unburdened, and not just of the downward pressure of gravity, but of the downward pressure of being; an intimation of what it might be like to let go of everything?

2 | Are we to lose our way amongst the immense indifference of things? Size is what the universe does. Nature iterates. Cut a sheet of paper in half, cut that half in half, and repeat the process thirty times. The piece of paper is now the size of an atom; except that, of course, as the size of an atom it would no longer be a piece of paper but an atom from which the paper was constructed; all the qualities that made it paper have disappeared. A further seventeen cuts to the size of a proton, and ninety-seven more to reach the smallest possible length that still has meaning: quantum length, $10^{-32}$ cm.* The original sheet doubled in size ninety times is as wide as the visible universe.

The more the universe seems comprehensible, the more it also seems pointless.

*Steven Weinberg, theoretical physicist*

---

* Mashing together quantum mechanics and general relativity suggests that perhaps even space and time come in smallest amounts. The smallest interval of space is a cube with side-length $10^{-33}$ cm, and the smallest interval of time – the time it takes light to travel across such a cube – is $10^{-44}$ seconds. Quantities of space and time smaller than these amounts are meaningless.

Not for us and not by the gods was this world made; there's too much wrong with it.

*Lucretius (c.99–55 BC), Roman poet and philosopher*

3 | 'I accept the universe' was a favourite phrase of the transcendentalist Margaret Fuller. 'Gad! She'd better,' said Thomas Carlyle.

4 | There are those who are emphatic that the universe has no purpose, but I do not know how they know. The scientific method doesn't do purpose, so it can hardly come as a surprise that the universe science describes is purposeless. What it all means is not a story scientists can tell. Science attempts to answer the question 'What is it like?' Not the question 'What is it?' If the point is what you are after, look elsewhere, not to science.

5 | Science describes a universe full of meaning. If there was no meaning there could be no science, and yet science has nothing to say about what that meaning means. We can know the facts, but why they mean anything is outside the remit of the scientific method. What if the world is evil? What if the world is a world of love? These are other kinds of investigations that human beings may undertake, but not when they are in the laboratory.

6 | Are the stars indifferent as they toil to make the ingredients necessary for life? Or is their seeming indifference an inevitable consequence of the way we tell stories and the way we do science: lineally from start to finish? In any case, our cosmologies are no longer stories of love, nor stories of intention. For those we must look elsewhere. The stars may be indifferent, but it is not their job to care: that is our human task.

7 | Perhaps the universe is pointless, but set as a dogma even pointlessness can be turned into purpose of a kind.

8 | Why should we suppose the universe inhospitable? Why not hospitable, because we are here? If this patch of the multiverse is an oasis, and if the earth is an oasis within this patch of the universe, well, we are in the oasis, and from this perspective – home – why should we be fearful of the desert? The universe is only inhospitable if we are against it. Why would we want to be against what has produced us, one of its most sophisticated products?

9 | The universe has the curious quality that it appears to be whatever we think it is: loving to the loving, fearful to the fearful, a world of anger to the angry, depressing to the depressed. When you're smiling, when you're smiling, the whole world smiles with you. When you're laughing, when you're laughing, the sun comes shining through. Like the Oracle, the universe is only as powerful as the questions asked of it. What would the world look like if it were rational?, is what science asks. Clearly a powerful way of questioning the world; the material world we live in being continuing evidence of that power.

> There is nothing either good or bad, but thinking makes it so.
> *Hamlet*

> Any universe simple enough to be understood is too simple to produce a mind able to understand it.
> *John Barrow's First Law*

10 | The cosmologist Brian Greene has argued that if eternal inflation really did happen, then there must be an infinite number of copies of you somewhere out there.* The argument goes like this:

i The 'visible' universe – as far as we can see, using our best physical theories and best measuring instruments – circumscribes a fixed amount of stuff, and that stuff is made out of particles. Add up what is out there and the visible universe is fashioned out of some $10^{80}$ particles.

ii If space is infinite (we do not know if it is or it is not, but most cosmologists believe that it is), then there is an infinite number of regions that circumscribe $10^{80}$ particles.

iii Quantum mechanics puts a limit on how reality is measured, with the consequence that there is a finite number of possible states that any finite number of particles can take. And so, logically, if the universe is infinite, there must be an infinite number of universes like ours, made out of $10^{80}$ particles, because anything that is finite endlessly repeated within something infinite must also be infinite.

11 | Brian Greene says that he takes comfort from knowing that there are other selves living out all possible versions of his life in far-flung reaches of the universe. The cosmologist Lee Smolin, however, finds

---

* This may seem hard to believe. But historically, scientists who refused to believe in the physical reality even of their own inventions and discoveries have often come unstuck. Einstein was one of its architects, yet he refused to believe in the physical reality of quantum mechanics, the most finely tested of all physical theories. For a time he did not believe in the physical reality of Black Holes, first discovered as a mathematical solution to the equations that describe the general theory of relativity; equations he had laboured over for ten years. About Black Holes he changed his mind. Even after he discovered the chromosome, Thomas Morgan continued to doubt the material existence of genes. Until well into the twentieth century the physical reality of the atom was questioned. Edwin Hubble didn't believe in an expanding universe, and yet his observations were the first evidence of such a theory.

such an idea horrifying, and wonders why he should care about the consequences of the choices he makes if all other possible choices (moral and immoral) are being made by an infinite number of Lee Smolins elsewhere.* I'm with Smolin.† I neither feel sorry for my other selves not making such good choices as I have made, nor envious of those that have made better ones. Why, anyway, would I choose to think of those universes that have me in them in some recognisable form and not all those other universes that do not, except to acknowledge that only the power of my imagination takes me to these far-flung parts of the multiverse?

> Of all the millions of ways Trim might have dropped his hat to signify the death of his master's nephew, there was this one and not the others.
>
> Laurence Sterne (1713–68), Tristram Shandy

12 | Trim doesn't exist, but he seems more real to me than the infinite number of copies of me out there in the multiverse, an infinite number of whom are writing this book and an infinite number of whom are not.

13 | Then there are other days: when I am willing to make the leap of faith that the mathematics offers and the physics requires, when I

---

* I am reminded of Nietzsche's idea of eternal recurrence. 'The universe must go through a calculable number of combinations in the great game of chance which constitutes its existence … In infinity, at some moment or other, every possible combination must once have been realised; not only this, but it must also have been realised an infinite number of times.' Even if our lives are not actually lived out repeatedly and eternally, we should live as if that were the case.

† Smolin does not hold with eternal inflation. He has proposed an alternative model of how universes are created. He calls it cosmological natural selection: 'Universes reproduce by the creation of new universes within black holes. Our universe is thus a descendant of another universe, born in one of its black holes, and every black hole in this universe is the seed of a new universe.' In Smolin's model our universe is typical of other universes of the same generation, whereas in eternal inflation our universe is a rare anomaly.

remember that cosmologists' flights of fancy call the scientific method to their defence, and accept that these other worlds do actually exist. On those days the flying carpets of imagination seem no more exotic, no less real, than the supposed realities of particle physics. Hurrah for a physical reality grown so cloudy.*

> You don't like it? Go somewhere else. To another universe, where the rules are simpler; philosophically more pleasing, more psychologically easy.
> *Richard Feynman*

> 'Imaginary' universes are so much more beautiful than this stupidly constructed 'real' one.
> *G.H. Hardy (1877–1947), mathematician*

14 | According to the theoretical biologist Stuart Kauffman, the world is more constrained than we suppose. Kauffman has called the constraints 'the adjacent possible'. Only certain next steps are available. Leibniz called it 'the compossible world'. Leibniz's insight was marred by his declaration that this world that we are experiencing is not only the only possible world, but the best of all possible worlds, a claim that was famously parodied by Voltaire in *Candide*.

15 | There are many things that are not possible in a particulate world. It is not possible, for example, to stand and run at the same time. The restraints on what forms the various island universes like ours can take may constrain them severely. The infinite may yet collapse into the few or even the one.

---

* The physical world is a hologram according to a theory first propounded in 1993. The world as it appears to us is a projection of a deeper reality that exists in more dimensions, just as Plato suspected. There is a self that exists in a higher-dimensional reality and moves as a shadow elsewhere, that elsewhere being what we call here.

For Macbeth's rhetoric about the impossibility of being many opposite things in the same moment, referred to the clumsy necessities of action and not to the subtler possibilities of feeling. We cannot speak a loyal word and be meanly silent; we cannot kill and not kill in the same moment; but a moment is wide enough for the loyal and mean desire, for the outlash of a murderous thought and the sharp backward stroke of repentance.

*George Eliot, Daniel Deronda*

16 | There is something circular about our physical descriptions of the universe. We suspect the universe might well be infinite beyond the horizon of the visible universe, and yet the laws of nature that underpin that understanding are derived precisely out of the limit of how far we can see. The finite speed of light, the fixed amount of time that has passed since the Big Bang, and the set number of particles out of which the contents of this region of the universe are fashioned obey laws that we only understand in the form that we understand them because we cannot see further than we do. If we could see further, we might discover, for instance, that the speed of light is not a constant after all.

Nor would the laws of physics as we understand them necessarily apply to universes with different numbers of particles in them. Nothing could be predicted if the universe had only a few particles in it. Even the so-called laws of nature have something statistical about them. The laws of nature, as we understand them to be, are what we uncover in a universe of this kind, with this many particles in it, seen from this perspective.

17 | We have forged our idea of reality out of what we can see. Nature crooks a finger and draws us on, and we follow in the hope of finding out things truer than those we knew before; revelation follows revelation, curtain after curtain is pulled aside – this, then this, then this – but there is no inner sanctum. There are days, most days, when I believe that the universe will not be outrun, not by the scientific method,

imagination, moral intuition, religious insight, nor any methods or combinations of methods of truth-seeking available to us.

We never get to the bottom of our understanding of the universe; there is always something more universal, more encompassing. We reach out to things truer but never arrive at the truth because there is no final destination. The starting conditions of the universe hold within them everything we don't know about the physical universe moved to the edges of time and space.

18 | It requires a certain kind of stubbornness not to see, in the fine measurements that science makes, confirmation of the existence of an external reality. But there are days when I find that I am that stubborn. How can we be separate from reality? Reach out and you reach into the Big Bang everywhere about you. The universe and we its observers have grown up together from the one stuff. What could we be separate as?

# Evidence for the existence
of an external world

1 | That science works – it creates what we mean by progress – is proof and evidence enough for most of us that there is a world out there, separate from us and full of things that have separate existences. And then there is mathematics, the strongest evidence of all.

2 | For the philosopher René Descartes the only certainties were mathematics and theology.

> The mathematics appear to be there in the behaviour of physical things and not merely imposed by us.
> *Roger Penrose, mathematical physicist and philosopher*

3 | Roger Penrose believes that mathematics has real existence in a kind of Platonic world parallel – but somehow connected (by what mechanism is not known) – to our world of experience.

> I believe that mathematical reality lies outside us, that our function is to discover or observe it, and that the theorems which we prove, and which we describe grandiloquently as our 'creations', are simply our notes of our observations.
> *G.H. Hardy, A Mathematician's Apology*

> Pure mathematics ... seems to me a rock on which all idealism
> founders: 317 is a prime, not because we think it is, or because our
> minds are shaped in one way rather than another, but because it
> is, because mathematical reality is built that way.
> *Ibid.*

4 | If we can believe with G.H. Hardy, Roger Penrose and others that
mathematics is out there, and not intertwined with our own perspective on the world, then materialists may claim, as some do, that one day
we will be able to write down, in the language of mathematics, laws that
fully describe the physical world. Mathematics appears to be proof that
the world can be transcended. Processes are real because they can be
described by mathematics, which is itself real. Mathematics looks like
proof that there is an external world, and that science is an investigation of its nature and substance.

> Mathematics is the only religion that has proved itself a religion.*
> *F. de Sua, mathematician*

---

* Sometimes mathematical belief turns mystical. The astronomer Johannes Kepler
(1571–1630) was obsessed with the number of planets in the solar system. He tried to
create a model of the solar system that related the number of planets and their orbits
to the set of Platonic geometrical forms. But his information was incomplete. He could
not know that more planets were yet to be discovered. In the 1920s, the astrophysicist
Arthur Eddington was fascinated by the unchanging numbers – the gravitational
constant, the speed of light and the charge on the electron – that appear in the laws of
physics, and speculated how they might relate to each other. He believed that they are
evidence of some deep unification of nature. The ratio of the gravitational force
between an electron and a proton is $10^{40}$. The size of the observable universe divided by
the radius of an electron is also $10^{40}$. Eddington calculated the number of particles in
the visible universe as 'precisely:' 15,747,724,136,275,002,577,605,653,961,181,555,468,044,
717,914,527,116,709,366,231,435,076,185,631,031,296 protons, 'and the same number of
electrons'. Named the Eddington number, it is close to $10^{80}$ ($10^{40 \times 2}$). The recurrence of
the number $10^{40}$ led the physicist Paul Dirac to wonder: 'Might it not be that all the
present events correspond to properties of this large number [$10^{40}$], and, more generally, that the whole history of the universe corresponds to properties of the whole
sequence of natural numbers ...? There is thus a possibility that the ancient dream of

The unreasonable effectiveness of mathematics.
*Unattributed*

The equation is smarter than I am.
*The theoretical physicist Paul Dirac (1902–84), on his equation*
*that predicted the existence of antiparticles*

5 | Deeper descriptions of the universe require more and more sophisticated mathematical formalisms. Einstein took ten years to find the mathematical language in which to write his general theory of relativity. Unusually, the mathematical formalism that quantum mechanics is written in came first, and its interpretation – still argued over – came afterwards.

It is only the unsophisticated outsider who imagines mathematicians make discoveries by turning the handle of some miraculous machine.
*G.H. Hardy, reminding us that mathematics is carried out by*
*human beings, not machines*

There is no sort of agreement about the nature of mathematical reality among either mathematicians or philosophers. Some hold that it is 'mental' and that in some sense we construct it, others that it is outside and independent of us. A man who could give a convincing account of mathematical reality would have solved very many of the most difficult problems of metaphysics. If he could include physical reality in his account, he would have solved them all.
*G.H. Hardy*

---

philosophers to connect all Nature with the properties of whole numbers will some day be realised.' Here in mathematics scientists and occultists meet: the world becomes some kind of code to be broken.

6 | Mathematical truths seem to have been out there waiting out the ages, to be discovered or not, by whoever or whatever.* It looks as if idealism founders on the rocks of mathematics. Not quite. There are loopholes. The philosopher Immanuel Kant (1724–1804) argues that science is an exploration of the co-evolution of humans and the universe; the doll and the dolls' house are inextricably entangled. Science is an exploration of that entanglement. Even mathematics, Kant believed, is not outside us: it is as much in our brains as in the outside world; yet it is less a mental construct than a product of the co-evolution of everything together, there being no meaningful separation between inside and outside. For Kant science is access to one kind of truth, another path being our sense of morality.

> Two things fill the heart with renewed and increasing awe and reverence the more often and the more steadily that they are meditated on: the starry skies above me and the moral law inside me.
> *Immanuel Kant, Critique of Practical Reason*

7 | Roger Penrose writes that mathematical models describe reality with 'a precision enormously exceeding that of any description free of mathematics'. Clearly this is true but somewhat circular. Physics does

---

\* Even the simplest-looking mathematical theorems may have to wait centuries for proof. In 1637 the amateur mathematician Pierre de Fermat claimed that he had a proof to show that there are no integer solutions to the equation $x^n+y^n=z^n$ for any n larger than 2. He made the claim in the margin of his copy of Aristotle's *Arithmetica*, adding that there was not space enough in the margin to give the details. No such proof has come to light and it seems likely that it was either flawed or did not exist. Andrew Wiles's proof of 1995 was seven years in the making and ran to a hundred pages. It was two years before other mathematicians felt confident enough in his workings to verify it. There are mathematical objects that are used in calculations even though they cannot be expressed. So where are they? Graham's number is the largest ever used in a mathematical calculation, but no one knows what the number is, how many digits it has or even what its first digit is; but it is known that it ends with a 7. The number was first described by a former circus performer named Ron Graham.

precision, poetry does metaphor. They are incommensurate. Biology cares for decimal points only somewhat; poetry not at all.

> Neither physicists nor philosophers have ever given any convincing account of what 'physical reality' is, or of how the physicist passes, from the confused mass of fact or sensation with which he starts, to the construction of the objects which he calls 'real'. Thus we cannot be said to know what the subject matter of physics is; but this need not prevent us from understanding roughly what a physicist is trying to do. It is plain that he is trying to correlate the incoherent body of crude fact confronting him with some definite and orderly scheme of abstract relations, the kind of scheme he can borrow only from mathematics.
>
> A mathematician, on the other hand, is working with his own mathematical reality.
>
> *G.H. Hardy*

8 | Einstein called mathematics the poetry of logical ideas.

9 | In the 1920s an attempt was made to make mathematical logic the sole means of advancing philosophical knowledge, and so rid the scientific method of metaphysics once and for all. The methodology was called logical positivism, and assumed experience as the sole source of knowledge. Logical positivists wanted to admit as meaningful only those sentences that can be independently verified, ultimately by mathematics.

> Logical positivism is the idea that a sentence or another fragment – something you can put in a computer file – means something in a freestanding way that doesn't require invoking the subjectivity of a human reader. Or, to put it in nerd-speak: 'The meaning of a sentence is the instruction to verify it.'
>
> *Jaron Lanier, computer scientist and virtual-reality pioneer*

Logical positivism is a form of solipsism. If you say physics is only about predicting the outcomes of experiments, you can only really say it's about experiments that you personally do, because to you any other person is just another thing you're observing. But solipsism is a dead-end philosophy and when it comes to science it's a poison.

*David Deutsch*

10 | At the laboratory in Lagado, the capital city of the nation Balnibarbi, there is a great machine that manipulates all the words of the language of the people. Scientists hope to extract knowledge out of the random generation of words. Where three or four words are found together that might make part of a sentence, they are dictated and transcribed into a 'large Folio already collated, of broken sentences'. Out of this process it is 'intended to piece together, and out of [these] rich materials to give the World a complete Body of all Arts and Sciences …' At the Mathematical School in the same city, formulae are written in ink on wafers and eaten, in the belief that the ink with its message will eventually reach the brain, where the information will be processed. Other scientists have attempted to reduce the nation's language entirely to nouns, since in reality 'all things imaginable are but Nouns'. Instead of speaking, language is reduced to pointing at things. The disadvantage is that it means carrying around a large number of things to point at. The women of the country rebelled and sought 'the Liberty to speak with their Tongues, after the manner of their Ancestors; Such constant irreconcilable Enemies to Science are the Common People.'

The quotes are from *Gulliver's Travels* by Jonathan Swift (1667–1745).

11 | Some materialists believe that out of enough data, meaning is self-generated. And so logical positivism creeps back into fashion. Larry Page, one of the founders of Google, believes that the internet will come alive at some point. The futurist George Dyson believes that it already has. Other futurists, Ray Kurzweil notably, predict that a moment will

come when machines will outsmart humans. It is predicted that this singular event will occur in the twenty-first century.

12 | Kurt Gödel's incompleteness theorem of 1931 (published when he was only twenty-five years old) presented the logical positivists' programme in a new light. Gödel's Theorem is often misstated. It does not show that mathematics is incomplete, but that mathematics is incomplete within any particular mathematical formalism, which is crucially different. It means, for example, that there are mathematical problems that can be written in the language of, for instance, arithmetic (an example of one kind of mathematical formalism), but cannot be proved in arithmetic. But arithmetic can be made complete within a more encompassing formalism, say, geometry; yet geometry is itself incomplete, and so on. Mathematics is a series of nesting formalisms, one inside another like a Russian doll. Our physical understanding of the universe – written as it is in mathematics – may be like this too.*

Gödel's theory is useful when we think about computers or any mechanical device that works in some formal way. It tells us that there

---

* Georg Cantor's proof that there is an infinite nest of infinities, each bigger than the last, is a similarly mind-melting concept. The infinity of decimal numbers, for example, is larger than the infinity of counting numbers. Further, there are always larger infinities than any particular infinity. The proof is so counterintuitive it made the philosopher Ludwig Wittgenstein furious, and still raises the blood pressure of philosophers and mathematicians. It raises the question, if the universe is infinite, what kind of infinite? Aristotle did not believe that infinity could be a quality of the earthly world down below. If he had, he might well have got to the principle of inertia two thousand years ahead of Newton. (For Aristotle things only moved if pushed; for Newton things remain in motion forever unless there is friction – the principle of inertia.) But Aristotle effectively ruled out the concept of inertia in order to exclude infinity. The Israeli mathematician Doron Zeilberger believes that there is a biggest number, and if you add 1 to it you return to 1. He belongs to a group of mathematical believers called ultrafinitists. He doesn't believe that the countable numbers extend to infinity. Cantor suffered a series of breakdowns from his forties until his death aged seventy-three. Gödel became mentally unstable in later life. He feared that he was being poisoned, and would only eat food that had been prepared by his wife. His wife became ill and was hospitalised. Gödel died of starvation.

are mathematical truths that cannot be proved by any computer or mechanical device, no matter how sophisticated that device may be. If our minds are formal in the way computers are, then there will always be some mathematics that is beyond our reach, and so also some understanding of the physical world that is beyond our reach. Alternatively, as Gödel pointed out, if humans can always delve deeper into mathematical reality, then they cannot be machines. We are left with two possibilities: humans are machines and their understanding of the world has a limit, or they are not machines and are free to explore the physical world forever. Either way the world is more mysterious than we are.

If we accept that mathematics is the strongest evidence we have of an externally existing world, Gödel's theorem throws a spanner in the works when we move on to consider the manufacture of our human doll. If human beings are not machines, then how can we make one? On the other hand, if they are machines, Gödel's theorem casts doubt on the possibility that humans themselves could ever construct a machine of equal complexity. We must hope that out of the scientific method we can create machines that are intelligent enough to evolve, and in that way eventually transcend the intelligence of their makers.

# SECTION 8

# Evidence against the existence of an external world

1 | Until the 1920s scientists believed in an independent reality that could be measured. But then from the 1920s there was quantum mechanics.

> I think I can safely say that nobody understands quantum mechanics.
> *Richard Feynman*

2 | The physicist Niels Bohr once said that quantum mechanics only makes sense if you change the meaning of the word 'understand'.

3 | Quantum mechanics tells us that the world is best described by a wave of superimposed probabilities. The most famous wave formulation is the Schrödinger equation. It is a linear superposition of different states of reality that evolves smoothly in time. Each possible observed reality has a certain probability attached to it.

4 | From the perspective of a molecule, nature is a single quantum wave of probability. From the perspective of a human being there are separate things, and particular events occurring at particular moments. The fundamental problem of quantum mechanics is how a reality that is described by a smooth wave also describes the world that we witness at our human scale. How are these two different perspectives commensurate?

5 | Various mechanisms have been put forward to explain how the sinuousness of the quantum world, in which everything is entangled in probability, becomes a world of separate things that actually happen, or appear to happen (the appearance being what we take to be the actual), but each is problematic.

6 | The Copenhagen interpretation, the most famous interpretation of quantum mechanics, devised largely by Niels Bohr and Werner Heisenberg in the 1920s, argued that quantum mechanics does not describe a physical reality but probabilities attached to the act of measuring itself. The wave of possibility collapses when a measurement is made, and we find ourselves in that world out of the many possible worlds contained within the wave. All the other possible worlds disappear.

7 | Light is both wave and particle. Light is made out of photons that in particle physics are understood to be particles. But light diffracts when it passes around the edges of a slit (or more evidently when it passes through a double slit). Light behaves as if it is a wave: troughs of light coincide to make the bands of darkness of a diffraction pattern. Even when the light is thinned so that single photon particles pass through the slits one at a time the diffraction pattern remains. It is as if each particle knows that it is part of the pattern. But how can a single particle know where to go? Follow the particle with a detector to see exactly when and where it decides which slit to pass through and the diffraction pattern disappears. The Copenhagen interpretation of quantum mechanics claims that it is the measurement itself that changes reality.

8 | The Copenhagen interpretation raises problems about what defines a measurement. It implies that the act of looking, of conscious attention, affects the outcome of an event. The attention of the observer creates the very thing we call reality.* But how much attention, and the

---

* Reality from our human perspective, that is. The 'realer' reality to a reductionist is the one quantum mechanics describes.

attention of what? As Einstein once asked, would a sidelong glance from a mouse suffice?

9 | The problem of how to interpret quantum mechanics became mixed up with the problem of how to define consciousness. If humans are uniquely conscious, as the Copenhagen interpretation seems to claim, then humans become highly privileged, a problem if the Copernican principle is to be upheld.

With human consciousness the universe became aware of itself. It might even be claimed, as the neurophysiologist John Eccles once did, that the universe cannot be said to have existed until there was human consciousness; the universe's past falling into place only retrospectively. What claim could be more arrogant, or anti-Copernican? Yet if we see reality as a conversation between human consciousness and what we take to be outside ourselves, the arrogance fades away. Other forms of consciousness have their own conversations with the universe.

10 | For Einstein, quantum mechanics only makes sense if there is some hidden variable which, once found, turns the world back into a world of cause and effect.

> Without being aware of it ... we exclude the subject of cognizance from the domain of nature that we endeavour to understand. We step with our person back into the part of an onlooker who does not belong to the world, which by this very procedure becomes an objective world.
> *The physicist Erwin Schrödinger (1887–1961) refutes Einstein's belief in an independent reality*

11 | If an observing mouse is a separate physical system, where does the system begin and end? With a few atoms it is clear, because atoms have lost their thingness. Out of pure thought alone, we might guess that a world of elementary particles has to be indistinguishable as things. How would elementary particles know where the boundary of the system is?

By the time we reduce the world to atoms, the world has become a place of limited sameness. The molecules of our bodies are continually in motion in and out across our boundaries, they do not know where we start and end. As complex structures of a certain size, humans lose their grip on reality and fall into the illusion of a world of things.

> Nature opens its eyes and sees that it exists.
> *Friedrich Schelling (1775–1854), philosopher*

12 | On the radio I hear Simon Saunders, a philosopher of physics, attempting to explain the measurement problem: Look, there's a fairly straightforward dilemma here, well trilemma really ...

13 | In Hugh Everett's 'many worlds' interpretation of 1957, reality is a single wave function, a superposition that never collapses of every state of everything in the entire universe. In the quantum world there is a rule: whatever can happen, happens. The many worlds interpretation of quantum mechanics says that all possibilities contained within the wave equation actually do happen. God does not play dice, all possibilities exist.

We see only part of the wave equation. Other selves see other parts of the wave equation in other parts of the universe. The idea of a coherent self is undermined. What we take to be our self is the illusion that arises out of being trapped in a particular perspective of the multiverse. Somewhere in the multiverse we are doing everything that it is possible for us to do.

14 | Nature is a single superposition. It is only with great effort that scientists create a rival superposition, a rival nature. At any instant Nature pounces to claim its own. The information of our rival laboratory-made superpositions will easily leak into the much larger superposition that is all of nature outside the laboratory. The leaking of this information gives the illusion that the wave collapses, but it is only an illusion: the larger wave absorbs the smaller wave.

Scientists have found ways to keep a molecule made of some twenty carbon atoms in its own superposition (outside the superposition that is Nature). It requires the creation of an extremely cold environment. The small molecule does indeed, under these extreme conditions, reveal its 'true' nature: being in two or more places at the same time, for example – behaviour that would have counted as magic less than a hundred years ago.*

We are far from working out how to keep something as complex as a cat separate from the superposition that is Nature. I once heard a scientist say that if we look hard enough, one day we might find cats and chairs embedded in the wave equation. Perhaps, but somehow I doubt it.

15 | The chance that all the molecules that make up a glass of water being together as a glass of water in some other part of the universe and not here cancel out all the possibilities of where it could also be, and it ends up being where we most likely take it to be, what we call 'here'. The rules of quantum mechanics tell us that all possibilities exist, but added together the unlikely ones cancel each other out, leaving what we acknowledge as the possible, the world of things. It is the addition of these possibilities that gives the illusion of a single reality, of a glass of water that exists very close to where we find it.

16 | Heisenberg said that the meaning of quantum mechanics is in the equations. Bohr, his mentor, pointed out that we still have to talk about the equation in words.

---

* Much of what is now possible in the material world would have seemed like magic in the past. Our machines are becoming more and more mysterious to us. There is no one alive today who understands the workings of all the machines that are now in the world. Our understanding of the world is held collectively. Arthur C. Clarke's Third Law: Any sufficiently advanced technology is indistinguishable from magic.

Mathematics is a great tool, but the ultimate governing language of science is language.

*Lee Smolin*

17 | The theoretical physicist David Bohm said that the quantum world is process based. Such a world is most readily described in verbs rather than nouns. He said that the problem with European languages, as in sentences like 'The cat sat on the mat', is that we are dealing with well-defined nouns, a world of things. He believed the process that the verb describes is the deeper reality from which the illusion of thingness emerged, not the other way about. It has been suggested that the Algonquin languages of the Blackfoot, Micmac and Ojibwa tribes are particularly suitable processed-based languages in which to see the world as it is described by quantum mechanics. For these tribes, the world is alive. Singing is generative. The world was sung into existence.

Explanations of processes by which things come to be produce a feeble impression compared with the mystery that lies under the process.

*T.S. Eliot (1888–1965), poet and essayist*

18 | There are physicists who are still uncomfortable with the world quantum mechanics describes. Many attempts have been made to undermine quantum mechanics as a description of how reality actually is, but all attempts have failed. Pragmatically, scientists all agree that quantum mechanics works, and most don't worry about how it should be interpreted.

# On time

Time flies like an arrow, flies like a banana.
*Attributed to Groucho Marx*

1 | From our human perspective the universe evolved us over time and through the action of one cause on another, and yet philosophical investigations of the nature of time, and of cause and effect, cast a shadow of physical doubt even here.

2 | Aristotle reasoned that humans cannot escape the chain of cause and effect, and that any chain of cause and effect must eventually find its origin in a first cause. In order to avoid this kind of reasoning he came to the conclusion that the universe must be eternally existent. Some scientists were made uneasy by the Big Bang theory because it isolates a beginning moment for the universe. The multiverse, created out of eternal inflation, randomly and without cause, is an attempt to make the universe once more eternal and uncreated as Aristotle envisioned it. And yet it is a creation story still, and one that will eventually be succeeded by some more refined account.

3 | Out of many possibilities of the quantum world a particular occurrence happens in time. But what is this moment of time in which things happen? How long is a moment? What joins one moment to another?

At the conjuror's, we detect the hair by which he moves his puppet, but we have not eyes sharp enough to descry the thread that ties cause and effect.

*Ralph Waldo Emerson (1803–82), American transcendentalist*

4 | Where does a cause end and its effect begin? What surrounds the cause and cuts it off from what it effects? What is in the space in between? What is the time between time? These are thoughts a precocious, gloomy child might have. But not childish; the problems run deep. The Scottish philosopher David Hume (1711–76) tried to pin down the elements of the illusion as a list of propositions that must be fulfilled: the cause and effect must be contiguous in space and time, the cause must come before the effect, there must be constant union between cause and effect; there are five further, increasingly elaborate tenets. Hume's insight was to see that cause and effect are habits of the mind. We associate two events, two stimuli, two ideas in our minds, and time passes. But if an idea existed truly separate from another, how would we ever move from that idea to the next? We would be attached to the idea forever, unchanging and frozen out of time.

5 | The world is this then this, not this because of this. If there is intention in the universe it is hidden at all levels. Or does not exist.

The very idea of a cause is emergent and abstract. It is mentioned nowhere in the laws of motion of elementary particles and, as philosopher David Hume pointed out, we cannot perceive causation, only a succession of events.

*David Deutsch*

6 | In Dr Johnson mode, a harrumph is enough to dismiss the problem. I turn on the kettle, the element heats up, the water boils, a cup of tea is made. Causes and effects. We know what a cause is and what an effect is in our world of large things and from our human perspective. At the gross scale of human beings consciousness seeps in, granting us

among its manifest powers the power to manifest cause and effect. But from a universal perspective that removes the human, the world must look differently. Pure thought tells us that there can be no gap between cause and effect – what could be in the gap? – but if there is no gap then there is nothing to distinguish cause and effect. Reductive materialism must and does account for this seeming impossibility. Relativity and quantum mechanics are two such accounts. Philosophy is not entirely useless! In relativity and quantum mechanics any exact formulation of what a cause can be is abandoned. Cause and effect are not qualities of the world at quantum scales. At the Big Bang the whole universe was a quantum event. Causality drains from the universe as we rush back towards its beginning. Radioactivity is an effect without a cause; the emergence of the universe from nothing is another example. Cause and effect emerge at human scales.

7 | Newton's equations might suffice to retrace the path of a ball or of a planet, or predict their future courses, but the ball's path and the planet's orbit are not contained in the equations. As powerful as the predictions contained in the equations are, they always describe only some limited part of what is going on. The planet's orbit might be changed because of some comet not taken account of. The ball may be thrown off course because of some unforeseen gust of wind. As hard as we try to predict what elements of change may arise, to describe any system, completely, we must eventually take account of the whole universe.

All this seemed to have ended well for Evgeny Mikhailovich and the yard porter Vassily: but it only seemed so. Things happened which no one saw but which were more important than all that people did see.

Tolstoy, 'The Forged Coupon'

So, whatever the verdict of physics, the real causal explanation for why there are boiled eggs is that I, and other breakfasters, intend that boiled eggs should exist.

*Alfred Gell (1945–97), anthropologist*

8 | Time is an illusion, said the ancient Greek philosopher Parmenides, and the deeper reality is eternal and unchanging.

Now he has departed from this strange world a little ahead of me. That signifies nothing. For us believing physicists the distinction between past, present, and future is only a stubbornly persistent illusion.

*Einstein writing, a month before his own death, about the recent death of his lifelong friend Michele Besso*

9 | For Einstein the past and future exist eternally. Time does not flow, it just is. In fact all physical theories so far devised – Newtonian mechanics, Einstein's two theories of relativity, even quantum mechanics – are symmetric in time. No arrow of time is indicated. There is no physical reason why a smashed plate might not rejoin itself, and indeed in some parts of the universe we should expect to see the arrow of time reversed. So far no such evidence has been found. To a material reductionist the arrow of time is an illusion of scale, just as Einstein's special theory of relativity shows us that it is an illusion that time and space are separate. It is because humans experience the arrow of time that the second law of thermodynamics* was added to physics. By fiat, the second law gives time a direction. The second law of thermodynamics fulfils an observational and psychological need rather than a physical one.

---

* Any isolated system evolves towards equilibrium, i.e. becomes as disordered as it can be. However, as Lee Smolin has observed, the universe itself is nowhere in equilibrium. The universe appears to be evolving forever into novelty. See page 67, 16.

The objective world simply is, it does not happen. Only to the gaze of my consciousness, crawling upward along the world line of my body, does a section of the world come to life as a fleeting image in space which continuously changes in time.
*Hermann Weyl (1885–1955), mathematician*

The physical world does not have tensed time, in which present, past and future exist side by side.
*Raymond Tallis, philosopher*

10 | Our human experience of living in the world is of time running forwards. We see a world in which everything, eventually, is ruined by time moving inexorably from past to future via a privileged instant we call now. The past is what the future becomes when it has been pulled through the ring of the present, and the present is the flame that burns the future into the ash of the past. Physical theories do not privilege the 'now'. Physics tells us that everything that will ever happen in the universe has already happened. The universe simply is. If material reductionists are to hold fast to their theories and the God-like perspective of physics, they must explain why we human beings experience the illusion of an arrow of time, and why the moments of our lives cannot be revisited. Why human beings are consigned to march, once only, second by second, forwards along a line of allotted time is a question so far unanswered by science. Or rather there is no consensus around whatever theories have been put forward.

11 | That there is no dedicated sensory organ that detects time\* might suggest that the passing of time is a psychological phenomenon.

There is no mechanism to go wrong. We can have a fragmented sense of self, but no one has ever had a fragmented sense of time. Our subjective perception of it, however, causes time to tick variously. Physical time ticks regularly, subjective time ticks fast or slow depending on our age, the emotions we experience, whether we are in pain, or

---

\* Nor is there any that detects space.

in love, or just bored. Dostoevsky writes of the condemned man's last night in which each moment stretches into eternity. Is that why we fear death: because at the last, the moment of death never quite ends?*

To think is to be in time. What we cannot do is think ourselves into the future. Our inability to find the future except by waiting for it to arrive in time suggests that consciousness itself is based in time.†

> Oh! Do not attack me with your watch. A watch is always either too fast or too slow. I cannot be dictated to by a watch.
> *Mary Crawford, in Jane Austen's Mansfield Park*

> If we knew how long a night or a day was to a child, we might understand a great deal more about childhood … It may be that, subjectively, a childhood is at least equal in length to the rest of a lifetime.
> *John Berger, A Fortunate Man*

> Einstein said that the problem of the Now worried him seriously. He explained that the experience of the Now means something special for man, something essentially different from the past and the future, but that this important difference does not and cannot occur within physics. That this experience cannot be grasped by science seemed to him a matter of painful but inevitable regret.
> *Rudolf Carnap (1891–1970), philosopher, reporting a conversation he had with Einstein*

---

\* There may be as many as 80,000 human beings in solitary confinement in the United States, a figure that is difficult to assess precisely, but has increased over time. In 2012 two men in Louisiana became the first humans in recorded American history to spend forty years in solitary. Each spends twenty-three hours out of every twenty-four in a cell that measures six feet by nine.

† People used to slip through time. Do they still? Sometime in the 1880s Miss Moberley and Miss Jourdain were walking about the gardens of the Petit Trianon at Versailles: turning a corner, they found themselves in the court of Louis XVI. *Kairos* is an ancient Greek word for the sensation of falling through time and finding yourself in eternity.

12 | The physicist Julian Barbour uses quantum mechanics to explain the illusion of the forward movement of time. The universe is a heap of frozen moments that exist eternally. Each of these moments is some quantum configuration of the universe. One moment may appear to us to come after another, but it is only because the second moment contains a memory of the first moment. The passing of time is an illusion. The second moment does not come after the first moment in time (time does not exist). The first moment is not the cause of the second moment (there are no causes).

We do not die in time, there is for each of us a small heap of frozen moments that is eternally our life.

Barbour's theory shows how the probabilities of quantum mechanics can be interpreted physically. Some of the configurations of the universe are much more likely than others. Time is the illusory ordering of these groups of more and more likely configurations.

13 | The arrow of time describes a universe that becomes more and more disordered, which means that looking backwards in time the universe must have been more and more ordered in its past. And so the arrow of time predicts that at its beginning – if it had a beginning – the universe must have been as ordered as it could be. But we also believe that the universe was very hot at the Big Bang – around $10^{32}$ degrees. Hot usually implies disorder. How the universe could have been so hot and so ordered is a mystery.*

14 | Our only human access to the past is via atoms. Becoming smaller and smaller, we leave everything behind. All the edifices, all the things of the world crumble into atoms, into electrons and protons as we spiral down towards the Big Bang. They are us but not we.

---

* Because temperature is a measure of the average motion of atoms, it has been suggested that the unity of the world at the Big Bang makes the idea of temperature meaningless at that moment.

15 | i Given the means, we humans could travel as far into the future as we care to. Come the day that we work out how to travel at speeds close to the speed of light we will for the first time know from experience, rather than theory only, what it is to leave earth-home. For now, because we live here together on the same planet we are united under a single clock.

I imagine myself travelling for long ages and at great velocity out into the galaxy. I have to learn to bear my homesickness. One day I decide to return home. The more I accelerate to hurry back, the greater the aeons of time that will pass on earth as I draw near. The lives of generations of human beings will have been breathed in and out, bones long crumbled into dust, reabsorbed into the biosphere and refashioned. Ice caps melt and refreeze, mountains lower and rise. The Himalayas have been a work in progress for fifty million years, and still they continue to rise; but even they one day will have been rubbed or convulsed once more out of existence if I travel far enough forwards in time. And I will return to what? Perhaps to some brutalised civilisation that has lost the past knowledge of long-forgotten golden ages. Or if I am fortunate, I might accidentally land in some new golden age of elevated beings beyond my imagining. If Julius Caesar* had had the power I have been granted, we might see him return tomorrow, full of amazement at the world he left behind. With the right means, we might travel as far forward in time as we wish, but only at the expense of leaving behind forever those who do not travel with us. For the first time in human history our human-scale clocks will not all tick together, here, where we live huddled together, in our living-room earth.

ii Children know how to make time machines. Put things in a box and bury the box. Wait.

iii No Penelope awaits her time-travelling Odysseus. When this Odysseus returns home the palace walls have long since fallen. The dog

---

* At his death it was recorded that he ascended into the sky.

that waited out its life died centuries before. Only the lead-lined box survives, carefully buried and mapped, which with shaking hands Odysseus opens.

iv There have been few human space travellers to date, but even amongst that small number a significant proportion have returned traumatised. Humans can go mad sailing alone across the oceans; how would even the most robust of us fare if we were to travel to other planets around other stars?

16 | For Lee Smolin the flow of time 'is not an illusion. It is the best clue we have to fundamental reality.' Smolin characterises the eternal laws of physics as 'excess metaphysical baggage'. He believes that the pursuit of eternal laws of nature has hampered cosmology in recent decades, and that mathematics is part of the problem – because mathematics looks like evidence of something eternal and unchanging in the universe.

In place of eternal inflation Smolin has posited a theory of cosmological natural selection to explain how the universe got going in time. He reconceives the universe as if it were some kind of organism. Everything in the universe is evolving at every scale into genuine novelty. In a universe in which everything is laid out eternally and unchanging, innovation is impossible. In Smolin's model every instant that happens in the present is a chance for something genuinely new to occur in the universe's history.

Smolin believes that at the quantum level the deeper reality is time, not space. He thinks that one day we will understand how space emerges out of some deeper order based in time.

> I think it likely that space will turn out to be an illusion ... a way
> to organise our impressions of things on a large scale but only a
> rough and emergent way to see the world as a whole.
> *Lee Smolin*

# SECTION 10

# On things

1 | Human beings like to make things, but when the universe makes things what are they? Being in the universe calls the thingness of things into doubt.

2 | i Our best scientific thinking tells us that reality is only as it seems to be because it rests on the foundation of a deeper reality in which separation and location are meaningless concepts.

ii What happens to radiation in an expanding universe is that it appears to become a landscape in which there are things. The appearance is compelling. The illusion of separate things is what the world looks like from our perspective, at human scale. That things are an illusion does not mean that they do not exist, but that they are not what they appear to be.

3 | Experiments impose a degree of artificial isolation, but nothing is truly isolated except perhaps in human imagination.

> He said that if we examine the various ideologies that tend to divide humanity, such as racism, extreme nationalism and the Marxist class struggle, one of the key factors of their origin is the

tendency to perceive things as inherently divided and disconnected. From this misconception springs the belief that each of these divisions is essentially independent and self-existent.

*The Dalai Lama, on a conversation he had with the physicist David Bohm (1917–92)*

In a mirror* we see from our reflection that there is no inside nor outside, and so 'things are freed from their thing-ness, their isolation, without being deprived of their form; they are divested of their materiality without being dissolved'.

*From the Atamsaka sutra, known as the mirror teaching, attributed to the Indian mystic Nāgārjuna of the second century AD.*

'Like' and 'like' and 'like' – but what is the thing that lies beneath the semblance of the thing?

*Virginia Woolf (1882–1941), The Waves*

We don't know what a rock really is, or an atom, or an electron. We can only observe how they interact with other things and thereby describe their relational properties.

*Lee Smolin*

The world is the totality of facts not of things.

*Ludwig Wittgenstein (1889–1951), philosopher*

Knowledge about a thing is not the thing itself.

*Henry James (1843–1916), writer*

---

* At school I was taught that a mirror reverses my handedness: left becomes right, and right left. This cannot be. I am the same person, my left hand is still my left hand even in the mirror. It only appears that I am looking at myself. I am standing just *behind* myself, a slice of myself a photon thick. I only appear to be looking back.

We seek the absolute everywhere, and only ever find things.
*Novalis (1772–1801), German poet and philosopher*

We are tormented with the opinion we have of things, and not by things themselves.
*Laurence Sterne, in Tristram Shandy, paraphrasing the ancient Greek philosopher Epictetus*

'Things' were of course the sum of the world; only, for Mrs Gereth, the sum of the world was rare French furniture and Oriental china. She could at a stretch imagine people's not having, but she couldn't imagine their not wanting and not missing.
*Henry James, The Spoils of Poynton*

Outside our consciousness there lies the cold and alien world of actual things.
*Heinrich Hertz (1857–1894), physicist\**

4 | Naming creates the world that is to be investigated. The world becomes a world of things, and sometimes, perilously, the world becomes a world of mere things.

5 | The universe exists as a physical object, the physicist Alan Guth claims.

6 | Because they are identical to each other, elementary particles are not things. Being identical, they can have no identity as things: that is what identical means. If there was any way of distinguishing one electron from another, that would be proof that they were not identical to each other. But if electrons cannot be distinguished one from the other, in what sense are they not the same electron? The physicist John

---

\* He was the first to prove by measurement the existence of electromagnetic waves. The phenomenon had been described theoretically by William Clerk Maxwell.

Wheeler wondered if all $10^{80}$ in the visible universe might not in fact be the same electron taking full advantage of that quality of the quantum world that allows a particle to be in more than one place at the same time. It is one of those ideas that feels so true that it must be true, yet Richard Feynman, when Wheeler told him of his idea, took just a few moments to prove mathematically why it cannot be true. Wheeler, however, may yet have the last laugh. A reformulation of general relativity called shape dynamics uses a similar idea. In shape dynamics, shape rather than size is used to relate objects far apart. The theory puts forward the idea that at its beginning the universe had a vast number of dimensions – a consequence of everything being much the same shape. As the universe evolved into greater complexity – more complexly shaped things – so the dimensions of the universe got pruned down to the three that complex humans are aware of. In shape dynamics, three-dimensional space is an illusion, and time the deeper reality that explains the illusion.

> For the first time in my life, then, I heard the voice of the One coming from the Many – I who until then had been taught to look for the wonder of infinite divisibility and variety, for the many in the one, the elaboration and detail of a broken infinity. My world, all through my life, had been made of parts ever increasingly divided into more intricate and complex fractions. By our contemplation of pieces of things we had grown to believe that the part is greater than the whole; and so division had motivated all the activities of people I had known, of books I had read, of music I had heard, and of pictures I had seen.
> *Mabel Dodge Luhan (1879–1962), New York heiress who set up a literary colony in Taos, New Mexico, and married a Native American, from her memoir Edge of Taos Desert*

'This must be the wood', she said thoughtfully to herself, 'where things have no names. I wonder what'll become of my name when I go in? I shouldn't like to lose it at all – because they'd have to give me another, and it would be almost certain to be an ugly one. But then the fun would be, trying to find the creature that had got my old name! That's just like the advertisements, you know, when people lose dogs – 'answers to the name of "Dash": had on a brass collar' – just fancy calling everything you met "Alice," till one of them answered! Only they wouldn't answer at all, if they were wise.'

She was rambling on in this way when she reached the wood: it looked very cool and shady. 'Well, at any rate it's a great comfort,' she said as she stepped under the trees, 'after being so hot, to get into the – into what?' she went on, rather surprised at not being able to think of the word. 'I mean to get under the – under the – under this, you know!' putting her hand on the trunk of the tree. 'What does it call itself, I wonder? I do believe it's got no name – why, to be sure it hasn't!'

She stood silent for a minute, thinking: then she suddenly began again. 'Then it really has happened, after all! And now, who am I? I will remember, if I can! I'm determined to do it!' But being determined didn't help much, and all she could say, after a great deal of puzzling, was, 'L, I know it begins with L!'

Just then a Fawn came wandering by: it looked at Alice with its large gentle eyes, but didn't seem at all frightened. 'Here then! Here then!' Alice said, as she held out her hand and tried to stroke it; but it only started back a little, and then stood looking at her again.

'What do you call yourself?' the Fawn said at last. Such a soft sweet voice it had!

'I wish I knew!' thought poor Alice. She answered, rather sadly, 'Nothing, just now.'

'Think again,' it said: 'that won't do.'

Alice thought, but nothing came of it. 'Please, would you tell me what you call yourself?' she said timidly. 'I think that might help a little.'

'I'll tell you, if you'll come a little further on,' the Fawn said. 'I can't remember here.'

So they walked on together through the wood, Alice with her arms clasped lovingly round the soft neck of the Fawn, till they came out into another open field, and here the Fawn gave a sudden bound into the air, and shook itself free from Alice's arm. 'I'm a Fawn!' it cried out in a voice of delight: 'And, dear me! You're a human child!' A sudden look of alarm came into its beautiful brown eyes, and in another moment it had darted away at full speed.
*Lewis Carroll (1832–98), Through the Looking-Glass*

Is love when you don't give a name to the identity of things?
*Clarice Lispector (1920–77), The Passion According to G.H.*

To feel simply that's a chair, that's a table, and yet at the same time, it's a miracle, it's an ecstasy.
*Virginia Woolf, To the Lighthouse*

7 | There are days when I know that the world is made out of things I can reach out and touch, but there are days, too, rarer days, when I know that everything I reach out to is one thing, and then, for a moment, I know that there are no things.

8 | We must learn to be separate. We were born inseparable from mother and the world. Counting, abstraction, these abilities had to be taught to us. And what we learned was to make a world as if of things.

9 | When King Midas was given the touch, the whole world turned to gold. The power recognised no boundaries between things. Midas breathed out and the air turned to gold. The atmosphere crashed to earth. The light that hit his retina turned to gold. All light and all radiation across the universe were transformed. The universe turned into a block of gold and fell to where?

10 | Atoms are continually on the move in and out of our bodies. Matter flows through us. From the perspective of the constituents of matter we have no boundary as a thing. We are part of a process that flows into everything. The boundaries emerge at different scales. The boundaries are illusions of scale and perspective.

11 | You look at an oak tree. You see its arbitrary boundaries in space: a trunk, roots, branches, twigs, leaves. You acknowledge the tree's arbitrary beginnings in time as an acorn. You think of all the tree's acorn ancestors, a line of acorns that stretches back in time to when the tree's precursors were no longer recognisably oaks, and then to when they were no longer recognisably trees, and so, inexorably, the origins of this tree draw you back to the origins of the universe (and back beyond to the abstractions of the multiverse of random quantum inflation, or whatever the latest modish physical theory might be). And then you trace this particular tree forwards in time from when it was an acorn, and see that it has been woven out of an iterative process of information-exchange between light, water, minerals and the code held in the seed. You see that the tree is all the sunlight that has ever fallen on it, light that has travelled across the solar system, a part of the sun captured and transformed here on earth into chemical bonds and radiant heat.* The tree is the water absorbed at its roots, it is all the water that has transpired from its leaves and been absorbed into the air, all the oxygen and carbon dioxide exchanged. It is the wind that has shaped it, and the parasites that live in its bark, that have mottled its leaves, the cankers and the mistletoe.† It is the birds that nest in it and the birds that have nested in it, even the birds and insects that have briefly landed on its leaves and branches. It is the cows that have rubbed against it, and the initials lovers have carved on its trunk. It is the roots that stretch

---

* As infrared radiation the photon will eventually escape the earth once more. At some point it will meet some obstacle – perhaps a mote of dust – deep in space, and so its eternal life continues.

† Mistletoe and oaks are culturally tied, but in fact mistletoe rarely grows on oak trees.

underground and join the roots of other trees. It is everyone who has ever sat in its shade. Whatever the tree is, it is you who decides where it begins and ends. There are no separate things, only what we make so. And so our investigations of the world begin as conscious human acts of separation made by observers. They begin in subjectivity, that, once the boundaries of our objects are decided upon, we call objectivity. You can say what a tree is and what it means to you, but from the point of view of the world, the world is of a piece and there is no tree.

And then snow descends and unifies the tree into the landscape. Snow unifies the suited tree to the field, even strident modern houses and factories and cars are reclaimed as natural forms, softened and smudged in. Gentle, muffling snow reminds us that the world is one.

The tree is as it is seen and painted by the artist. No one can draw a tree without in some measure becoming that tree, a painter once told Emerson. In a Gainsborough portrait the feathers in the woman's hat have the same featheriness as the leaves on the tree.

The sunlight measures the tree, as does the wind passing through its branches, and you absorb the sunlight and the sounds and you measure the tree to your own ends. Or at night you make out the tree in a different fashion out of starlight. Even without the sun and stars you might – in theory – be able to measure the tree out of its relation to the particles coming into and out of existence in the vacuum. If the vacuum were truly empty there would be no tree. Yet you are conscious of the tree as a tree, as a unified thing. This is a profound problem.

# SECTION 11

# Starting again

1 | Physicists' seeming obsession with fundamental descriptions takes them as far away from the human condition as it is possible to go, to the beginnings of the universe. Even though they often pretend otherwise – affecting to disdain philosophy – physicists care what the fundamentals are: what the material is that reality is woven out of; whether it is space that is real and time the illusion, or vice versa; whether or not there really are causes and effects; how the world can be all of a piece and yet have things in it. If we are ever to make a human being we will have to abandon these concerns and move closer to our own scale. In this evolving, expanding universe in which we find ourselves we might take everything to be equally existent no matter when it first appeared, or at what scale. Time may be an illusion from a reductive physical point of view, but it is no less real now at this point in the evolution of the universe. There are causes and effects in the human world. Things happen, and we make things. We might take the universe as read and wonder instead how the universe evolved human beings as things in time.

# PART TWO

# Animating the Doll

Who's there?

*The opening words of Hamlet*

# Matter → meat

Darwin gives courage to the rest of science that we shall end up
understanding literally everything, springing from almost nothing
– a thought extremely hard to comprehend and believe.
*Richard Dawkins*

1 | These days evolution is invoked to describe not only the evolution
of life, but the evolution of the whole universe. Evolution is the story of
what happened, in an expanding universe, to those first particles
created at the Big Bang.

There is grandeur in this view of life, with its several powers,
having been originally breathed into a few forms or into one; and
that, whilst this planet has gone cycling on according to the fixed
laws of gravity, from so simple a beginning endless forms most
beautiful and most wonderful have been, and are being, evolved.
*Charles Darwin, On the Origin of Species*

If superior creatures from space ever visit earth, the first question
they will ask, in order to assess the level of our civilisation, is 'Have
they discovered evolution yet?'
*Richard Dawkins*

2 | For reasons not entirely understood, a patch of energy escapes the
eternally inflating quantum landscape and expands into a universe of

space and time. The energy evolves into energy of different kinds. An array of different kinds of particles come into existence spontaneously and randomly out of nothing, among them quarks, gluons and photons. As the universe expands the particles evolve into other kinds of particles. Under gravity the first light atoms – hydrogen and helium mostly – are drawn together as quasars and stars. Gravity arranges the stars as galaxies and galaxy superclusters. The stars catch fire as great furnaces that transmute light atoms into heavier atoms. When some of the stars of the right size die and explode the first heavy atoms are spewed into space. Gravity acts on the light atoms and the newly forged heavier atoms to make a second generation of stars. Some of these stars of the right size explode and release even heavier atoms into space. And repeat. Simple compounds like water, carbon dioxide and some amino acids are also produced. How the simple compounds evolve into DNA is not yet known (it may never be known), but few doubt that evolution is how it happened. Human beings tell a plausible, if patchy, tale of how hydrogen was woven into flesh. It may be the same tale aliens tell. Assuming that aliens are made of flesh. And tell stories.

3 | Physicists used to criticise biology for being too descriptive and not testable. All that changed with the advent of genetics and molecular biology. Less than a century ago natural selection alone could not account for the complexity of living forms, but genetic theory turned natural selection into a rigorously testable theory.

4 | Life used to distinguish biology from physics. For centuries, biology was plagued with the notion of vitalism, a mysterious – perhaps even mystical – force that was invoked to account for the seemingly unbridgeable gap between living and non-living forms. In his book *What is Life?* (1944) Erwin Schrödinger wondered if life might emerge out of some simple physics of self-replication and a single molecule. His book influenced many biologists. Francis Crick swapped physics for biology when he read it.

Genetics killed vitalism. Once we could explain how genes determine biological function, vitalism was redundant. The golden age of molecular biology was ushered in. The discovery of the gene revealed the astonishing relatedness of all living forms. It came as a great shock to biologists to discover how much genetic information is conserved across all life forms. Our gene sequences are 95 per cent identical to those of yeast. The difference between yeast and a human being is largely a matter of organisation. We share 50 per cent of our genetic code with a banana. Sixty per cent of genes in humans that are associated with disease have a homologous gene – i.e. the same gene making the same proteins and doing a similar job – in the fly. The Krebs cycle – a vital part of the metabolic process – is common to all forms of life, from single-celled bacteria to flies and kangaroos. Within the Krebs cycle is a tiny epicycle that generates ATP, the chemical that powers muscle contraction and nerve impulse. 'Where there is life,' writes chemist P.W. Atkins, 'there is ATP.' These observations can be proclaimed either derisively or with wonder. Do we have more in common with flies and bananas than we have allowed ourselves to imagine, or do genes only take us so far? Or both?

5 | How did life on earth first emerge?

Nobody has a freakin' clue.
*Steen Rasmussen, physicist specialising in artificial life*

I cannot myself see just how we shall ever decide how life originated.
*Francis Crick*

If you study life and you're not flabbergasted by the hypothesis of evolution, you just haven't looked carefully enough. I think it's a tragedy that nobody walks into a class and expresses that wonder ... We have no evidence that life is not a miracle, i.e. a very low probability event. And we should give that one away and say we don't know the hell where it came from. If you want to believe in God at that point, it's as good a theory as any other, but no one will say that, and that is a pity. It all has to be in the running.

*From a conversation between a scientist and the author*

The universe has the curious property of making living beings think that its unusual properties are unsympathetic to the existence of life when in fact they are essential for it.

*John Barrow, theoretical physicist*

The more I study the universe and study the details of its architecture, the more evidence I find that the universe in some sense must have known we were coming.

*Freeman Dyson, theoretical physicist*

6 | If any of the constants of nature – the mass of the electron, say – were different, so would the whole universe be.

7 | The early solar system was a violent place not conducive to life as we know it. A disk of dust orbiting our sun accreted under gravity and multiple collisions into numerous objects larger than dust-sized. After many such collisions and over a long period of time these numerous objects came together as planets and asteroids. Once these large objects had settled into fixed orbits and got out of the way of each other, the solar system gradually became a place where collisions are rare events rather than the norm.

Life on earth would not be possible if the earth's orbit was a little more elliptical than it is. The earth would repeatedly travel outside the narrow zone in which life as we know it here is sustainable.

Life on earth was not possible until the earth's temperature stopped varying so wildly. It took most of the earth's 4,500-million-year history before this was the case. The gradual stabilisation of the amount of carbon dioxide (and other so-called greenhouse gases) in the atmosphere has played an important role in the stabilisation of the earth's temperature. (Through the interference of humans, that stabilisation is being tested.)

Life would not be possible without water. At some point in the earth's deep history the atmosphere begins to fill with water vapour. There is a day when it rains for the first time. Some of the oldest fossil remains are of the indentations of rain made in rocks discovered in India. They are at least three billion years old, though by then it had been raining for at least a billion years.

Life would not be possible if water did not have the unusual property of being less dense in its liquid state than in its solid state, otherwise the oceans would have frozen from the bottom up and killed all the first life of the seas.

A fine day is rare enough even in the Cotswolds, and certainly it is not the default state of the world anywhere very much. If it is an accident that the earth is conducive to life it is only because of innumerable accidents. Innumerable accidents begin to look suspiciously like no accident at all.

What we don't know is if these conditions, and many others, would rule out life, or only life as we currently understand it.

8 | The numbers game doesn't work for life. With our best current measuring devices we are beginning to find other solar systems with proto-earths, and so the probability of there being life elsewhere goes up. But the probability also goes down, because we have not yet found life. On all the proto earths discovered so far the conditions are hellish. For as long as we do not find life elsewhere, what we find instead are the ever-finer conditions needed for life here. The more closely we investigate the conditions that did not result in life elsewhere, the more finely tuned those conditions become for life as it appeared here. There is no evidence that life is not a miracle; or, in scientific language, a very

low probability event.* How you answer the question: Is there life else-where? is a matter of taste.†

> There is nothing that God hath established in a constant course of nature, and which therefore is done every day, but would seem a Miracle, and exercise our admiration, if it were done but once.
>
> *John Donne (1572–1631), metaphysical poet*

9 | Science doesn't do miracles: better to believe that there are many universes with many different conditions, and that we necessarily inhabit one island universe – ideally one of many island universes – in which the conditions for life are as they need to be.‡

Science is the question, what does a world without the mediation of the supernatural look like?, repeated over and over again. If science were to accept irreducible mystery, or irreducible complexity, that would be to give up and admit defeat. Science moves into the unknown with the assumption that there is always something more that can be found out. To allow our universe special properties is one way of bring-ing scientific investigation to an end. To allow humans a privileged perspective is another. If humans have anything of the God-like about them they move beyond the reach of the scientific method. Instead, try

---

* The chance of throwing 1,024 marbles, randomly one at a time, into a box divided into two, so that all of them land in the same half of the box, is 1 to $10^{290}$ against. (There are $10^{80}$ elementary particles in the universe.) The chance is not zero, but 'unlikely' hardly seems to cover it. Any event that is unlikely to happen in the lifetime of the universe we might call functionally miraculous. It is functionally miraculous that life emerged in the universe.

† Or belief.

‡ Science used to hold to a principle called Occam's razor: the faith that between differ-ing but equally powerful scientific descriptions the more parsimonious one is the truer (because more elegant, just as mathematicians talk of inelegant proofs, and labour to find more beautiful ones for the sake of beauty rather than of advancing knowledge). In order to protect the Copernican idea of mankind's lack of privilege, modern cosmo-logical theories have thrown Occam's razor out of the window.

to imagine a more encompassing universe in which the specialness disappears. And carry on. As science progresses, what the universe can be gets subtler and more mysterious. And as a by-product, the gods, God, and human beings become subtler too. If as scientists we were pragmatically to ignore what happens only once in the lifetime of the universe we would disavow ourselves and our planet home. Out of the scientific method, we humans never quite come into view. We can always look further out, but we humans seem always to be just over the horizon.

10 | Darwin wrote that 'life with its several powers, was breathed by the Creator into a few forms or into one'. But this was his public stance. In a private letter he guessed that life first arose 'in some warm little pond, with all sorts of ammonia and phosphoric salts, light, heat, electricity, &c. present'.

> Life transcends any attempt to describe it ... Life is ... planetary exuberance, a solar phenomenon ... matter gone wild ... a question the universe poses to itself in the form of a human being ... the representation, the 'presencing' of past chemistries ... the watery, membrane-bound encapsulation of space-time ... animals at play ... a marvel of invention for cooling and warming ... the transmutation of sunlight.
> *Various definitions of life from* What is Life? *by Lynn Margulis and Dorion Sagan*\*

11 | Life, according to the biochemist Frederick Gowland Hopkins (1861–1947), is 'the expression of a particular dynamic equilibrium which obtains in a polyphasic system'. In other words, if you are not moving you must be dead.

---

\* Quoted in (clearly a popular title) *What is Life?*, by Ed Regis.

12 | Crystals and snowflakes are products of complex organisation, but life is more than that. Candles burn, but life is more than consumption. Life is what metabolises substance into actions in the world. Life requires self-organisation, and is the result of natural selection.

13 | DNA and RNA* replicate only in the presence of many other proteins, which suggests that whatever first-replicating molecule there was on earth, it has long since disappeared. There is some evidence that a simpler self-replicating RNA molecule once existed, so it is not inconceivable that the search might eventually progress backwards to (slightly) simpler molecules like TNA, PNA or ANA, and from there to who knows where.

14 | Freeman Dyson wonders if the first ingredients of life have disappeared without trace (just as all evidence of the first stars made entirely out of hydrogen and helium – called Population III stars – has vanished).

15 | In a now notorious experiment conducted in 1953 by Stanley Miller, water, ammonia, hydrogen and methane were brewed together in a flask. After a week, twenty chemical compounds were detected, including amino acids. Many claims have been made for this experiment. There is no evidence that anything like this happened in nature, but Miller did show that out of a group of inert chemicals some of the basic molecules needed for life can be produced. The experiment does not show how we get from these relatively simple molecules to the extraordinary complexity of the DNA molecule.

16 | DNA is what life on earth has in common, but it does not follow that DNA is necessary for life to exist. For biologists like Stu Kauffman, DNA was a frozen accident, a chance happening that has persisted. Life elsewhere in the universe may have begun in other ways. In what ways

---

* Like DNA, a nucleic acid essential for life as we know it.

depends on how we define life or what we discover life to be in the future.

17 | Gradients are important in life. The first gradient may have been the electrochemical gradient found at hydrothermal vents, where the first single-celled life is thought to have emerged.

18 | Life became more complex when single-celled bacteria came together in colonies.

19 | Complex living forms are made out of the repeated division of a single cell. If there was not some graduated difference between one side of a cell and the other, cells would reproduce exact copies of themselves, rather than the slight differences that eventually ensure that there are cells with different functions making up the different parts of a body. There are not many means used in nature to ensure asymmetric division. Some organisms use the point where the sperm enters the egg as a point of differentiation.

# Making babies

1 | The controlled growth of a single cell makes a human being.* A single cell divides – the helical DNA strand in its nucleus is prised apart and copied – and two cells result. Two cells divide and there are four cells. Repeat forty-eight times and there are around fifteen million million cells, enough to make a human.

The cells become specialised. Genes within the nucleus are switched on or off to define what proteins are produced in the cell (though it is not known by what mechanism. What switches on the switch?†). Most cells retain all the information needed to make another organism, which is the secret of cloning. Dolly the sheep was cloned out of the breast cells of a 'mother' sheep.

2 | The nuclei of egg and sperm fuse at the two-egg stage. For the first six days the early embryo is a confederation of cells that gradually fuse together before becoming implanted in the uterus. Where the baby begins and ends only gradually emerges as the embryo develops. At what point an embryo can be considered to be a human being cannot be resolved by biology alone. When the baby is born into the world it must be cut off from the placenta, which belongs as much to the mother as to the baby. Only then does the baby's journey of separation begin.

---

\* The uncontrolled growth of a single cell is the cancer that kills that same being.

† Similarly, what tells nerves to fire is a mystery.

3 | In the womb a baby gathers an average of 25,000 new neurons a minute over nine months, a hundred billion by full term, each connected to a thousand others in a trillion connections. In the foetal brain all parts are connected together. At birth the pruning begins. There are pruning genes that get to work: snip, snip. Interaction with the world cuts some connections and reinforces others. And so a unique person begins to be shaped. Between a fifth and a half of the connections a child is born with are lost during the first month of life. We are born with a highly connected brain that is sculpted from within, reinforcing some of the connections we are born with and removing many neurons altogether. We might suppose that children start out simply and become more complex, but perhaps the opposite is just as true.

4 | If a human body were to be separated into the individual cells that make it up, 90 per cent of them would turn out not to be human cells at all but bacterial ones. The human body hosts bacteria, some 100 trillion micro-organisms (between three hundred and a thousand different species) in the gut, trachea and mouth that help (mostly) the body to digest food. It is possible to live without bacteria, but life would be hampered.* Babies are sterile – bacteria-free – in the womb, but bacteria begin to invade the body immediately we are born, and we are never free of them again. A baby born vaginally acquires his first bacteria as he moves out into the world and comes into contact with his mother's faeces. Such a child is likely to have an established system of flora in his gut a month after birth, a baby born by Caesarian section six months after.

5 | Babies are born with inbuilt strategies. They are not blank slates. The neural mechanisms they possess are the productions of evolution. A newborn is on the lookout for face-like objects two minutes into the world; from five months old delighted when solid objects seem to

---

* Bacteria is, for example, necessary to produce vitamin K.

disappear. And the magic seems to stay with us. Who isn't still entertained by a disappearing trick?

6 | The child emerges from nothing, and for a time knows it. For a few the memory fades slowly enough that the child can remember remembering. The child shakes her younger brother and says, 'Quick tell me what it is like, I am forgetting.' In the wake of their birth babies trail other worlds. Do they cry for what is lost, or for what is gained?

> Our birth is but a sleep and a forgetting:
> The Soul that rises with us, our life's Star,
> Hath had elsewhere its setting,
> And cometh from afar:
> Not in entire forgetfulness,
> And not in utter nakedness,
> But trailing clouds of glory do we come
> *William Wordsworth (1770–1850), poet*

7 | Freud described the oceanic feeling, a memory of childhood when the ego was not separated from the world about us, 'a feeling of an indissoluble bond, of being with the external world as a whole.'*

8 | For an infant, time is in its infancy. As a baby there is no shoreline from which to cast off into time, and no experiences to order in time. We grow into time. Children and animals live in the present, adults live in time. Animals are not aware of the tenses of time. Animals have clock genes that tell them what time it is. They do not have a sense of time passing. They are synchronised to nature holistically.

9 | Children are egotistical because they are only recently individuated. Their selfishness pivots on two extremities: being unindividuated,

---

* 'Oceanic feeling' was a term first coined by the French writer Romain Rolland (1866–1944).

which is to be the whole world; and being a child, which is to be hardly anything at all.

10 | The child must be taught to differentiate herself from her mother and from the world itself. The child must unlearn what she is born knowing: that everything is one. Too fanciful? The child is not a child, but must be taught to be one. A child does not have a settled personality, she plays in order to construct the character she can play at being for life. A child's ability to shift attention is an ability actors must learn. Children have the longest period of immaturity of all animals. We teach the child that the world is fragmented, and the child is naturally distressed. We name the parts: look, a cow; here a flower; there a mouth. And then having separated things, we reunite them. We say: the flower is red, the fire engine is red, the dress is red. Two trees, two apples, two people, two houses. Do you see? And for a time the child does not. The state of no things is lost when the child makes the leap into abstraction, and into consciousness.

# SECTION 3

# On consciousness

1 | Meanwhile, far away from the beginning of the universe, we find ourselves looking on.

2 | In the trade the problem of consciousness is called, with what I take to be amused understatement, 'the hard problem'.

Tell me what is a thought, of what substance is it made?
*William Blake (1757–1827), poet*

There is only one sort of stuff, namely matter – the physical stuff of physics, chemistry and physiology – and the mind is somehow nothing but a physical phenomenon.
*Daniel Dennett*

The human brain is a machine which alone accounts for all our actions, our most private thoughts, our beliefs ... All our actions are products of the activity of our brains.
*Colin Blakemore, neurobiologist*

I regard the brain as a computer which will stop working when its components fail. There is no heaven or afterlife for broken down computers; that is a fairy story for people afraid of the dark.
*Stephen Hawking*

The position that makes the most sense to me is that one's physical
and mental characteristics are nothing but a manifestation of how
the particles in one's body are arranged. Specify the particle
arrangement and you've specified everything.
*Brian Greene*

3 | Materialists attempt to make the problem of consciousness disap-
pear. Consciousness is simply what arises out of the computational
workings of the brain. Consciousness emerges out of the complexity of
non-conscious parts working together.

4 | If the cell is the fundamental unit of life, the neuron looks like
being the fundamental unit of intelligence, learning, memory, and even
consciousness.

5 | The idea that the mind and the brain are the same thing is an
ancient idea that goes back at least to the ancient Greek physician
Hippocrates (c.460–c.370 BC).

6 | If the mind is simply what the brain does, then the problem of
consciousness is reduced to this: why are we are aware of some of the
brain's workings and not others?

7 | Consciousness: an awareness of being aware.

8 | Most of what goes on in the body is unconscious. Typing these
pages, my body is busy doing a million things I'm mostly not conscious
of, like digesting my lunch, breathing, keeping my body upright,
moving my fingers across the keys, and so on. There are fifty to a
hundred million neurons around my stomach arranged in a complex
network called the enteric nervous system, yet I have no consciousness
there. I do not know what the acidity of my stomach is, I do not know
how oxygenated my blood is.

9 | i According to an ancient tradition I have feeling knowledge of the internal workings of my body.

ii The four humors of the body – blood, yellow bile, black bile, and phlegm – were incorporated by Hippocrates into a medical theory of the human body. Over five hundred years later, the Roman physician Galen (AD 129–c.200) associated a temperament and an animal nature with each of the humors: sanguine (horse), choleric (lion), melancholic (sheep) and phlegmatic (pig).

iii Expressions preserved like fossils in our language hint at this unconscious knowledge of our bodies' internal workings: the sinking feeling in the stomach, the heartache that is love and not angina, what is felt in the bones or stirs in our waters.

There is more reason in your body than in your best wisdom.
*Friedrich Nietzsche (1844–1900), philosopher*

Inside the body, too, there is profound darkness, yet the blood reaches the heart, the brain is sightless yet can see, it is deaf yet hears, it has no hands yet reaches out. Clearly man is trapped in his own labyrinth.
*José Saramago (1922–2010), The Year of the Death of Ricardo Reis*

10 | The neuroscientist Antonio Damasio describes the mind as embodied, not just embrained. The philosopher Alva Nöe describes the mind as 'a world-involving dynamic process'. It would be hard to say what a brain in a vat could be. A brain only makes sense when it is attached to organs.

We take our skin to be the border between ourselves and a hostile world outside. The ego is what protects us from that hostile world. But the borders – physical or psychical – are illusory. The mind is not restricted to brain, but flows out through the nervous system into the whole body and from the body into the world. Consciousness in this scenario arises out of the interaction between the body and its environment. Because science has, historically, taken the physical world to be out there separate from us, it is not surprising that it has not succeeded in describing consciousness. The outside world cannot be separated from the experience we have of it.

11 | The conscious self is guided as if it were blind by the intelligence of the unconscious body, and the rational structures of the external world. Walking is hard to do, and requires complex neural events to take place. We are tottering creatures with massive heads precariously balanced on thin necks and weak spines, and yet (once we have learned to) we (most of the time) walk without conscious effort. We might guess that we will eventually know what set of operations is carried out by the brain to make the legs walk, but what set of operations carried out by the brain makes a mind? Is this question even coherent?

12 | If we insist that the conscious self arises out of machine-like brain processes alone, then, somehow, out of all the thoughts the brain constructs, we must explain how consciousness is attached to some thoughts and not to others. Why do we experience a tree, and not neurons firing? If such a mechanism is ever discovered, then the self is reduced to a privileged bundle of thoughts. But there is a profound difficulty with this approach. How can there be such a mechanism at all? What would it mean to attach privilege to one thought and not another? It implies that there is another self that is making the choice, and so we conscious beings collapse into an infinite regression of selves. Who controls the controller? – a kind of Zen koan* that might shock us into the understanding that there is no one at home.

---

* Koans are statements or questions, meant – if considered deeply and long enough – to startle the student into an understanding that separation (in whatever form) is an illusion. The most famous example is the question, What is the sound of one hand clapping? My friend S is currently contemplating the koan, How do you stop the sound of a temple bell?

13 | Consciousness comes in various states. Brain dead,* vegetative, coma, drink- and drug-induced, hynoptic, sleep, dreaming and non-dreaming, lucid dreaming, hallucinatory, hypnogogic,† hypnopompic,‡ trance, bliss, awake, fully awake. We know the bottom state is death, but we do not know measurably how awake it is possible to be. Do drugs widen consciousness and allow more of the world in, or is the change restricted to the brain? Is consciousness itself evolving as humans evolve through time?

14 | Consciousness appears to be a construction like virtual reality. In virtual worlds we lose the sense of our physical body and accept the avatar as our self. It's an illusion, but no less real for that. Illusions trick us into thinking the world seems to be other than it is. There is still a real world, but its true nature is hidden behind the illusion. It may be that we are destined to pull back curtains eternally on backdrops that turn out to be just another set of curtains. What we call real is the illusion we most believe in.

> Maybe if you wiggle your toes, the clouds in the sky will wiggle too. Then the clouds would start to feel like part of your body … The body and the rest of reality no longer have a prescribed boundary. So what are you at this point? You're floating in there, on a center of experience. You notice you exist, because what else could be going on? I think of VR [virtual reality] as a conscious-ness-noticing machine.
>
> *Jaron Lanier*

---

* The brain dead are mentally dead, not physically dead. A brain-dead pregnant patient was kept alive for 107 days so that her foetus could develop fully.

† Between waking and sleeping.

‡ Between sleeping and waking. During both hypnogogic and hypnopompic states the right-hand brain becomes dominant. It has been suggested that these states may explain alien-abduction stories.

15 | If we were ever able to make a seamless connection between our nervous systems and computers, would we finally lose our sense of separation from the world?

16 | It has been suggested that consciousness is like an eddy in water or the pattern of a snowflake. Particular physical constraints grant specialness to certain kinds of thought. Or perhaps consciousness is a category error, like 'ascribing nutritional value to prime numbers'. The philosopher Daniel Dennett says consciousness is merely a linguistic confusion. The neuroscientist Christof Koch says, try telling someone suffering great pain that what they are feeling is a linguistic confusion; that their pain is of no greater significance than any of the unconscious workings of their body.

17 | Perhaps mindstuff is some fundamental feature of the universe, like space, time, energy, mass or charge.

18 | The neuroscientist and anaesthetist Stuart Hameroff does not think that mind emerges out of the increasing complexity of machine-like parts. He believes that there is some intrinsic quality about the universe that we do not yet understand, the mindstuff of the universe. Consciousness doesn't disappear when we sleep – we become conscious in a different way – but when we are anaesthetised, consciousness disappears altogether. Anaesthetics dim consciousness from lights on to lights out, but we do not know how they do it. Hameroff believes that anaesthesia works at the quantum level. He thinks that consciousness may be a quantum property too. He and the mathematician Roger Penrose, who have worked closely together, believe that consciousness may somehow be explained as patterns in the fundamental quantum granularity of space-time geometry. And if that weren't trippy enough, Penrose thinks that this is also the place where Platonic mathematical forms exist.

19 | The philosopher John Searle wonders if mind could be explained as some kind of field whose properties are currently unknown to us. The trouble is, we are a long way from knowing even what kind of properties to ascribe to such a field. We may be waiting for the Newton of our age to outline the field's properties.* A psychic field would be a radical step, as gravity was, and for now such explanations are popular among pseudo-scientists, new agers, and a few brave scientists willing to stick their necks out.

20 | The physiologist John Eccles and the philosopher Karl Popper together took a radical approach, and suggested that biology might need to turn itself into physics.† They posited the self as a field that interacts with and controls the brain.

21 | There are those who hold that reality is made out of the material of consciousness, not out of matter. Generally this is a view more likely to be held by philosophers and artists – Proust perhaps most famously – than by scientists, though there are notable exceptions. The belief that everything is ultimately the content of consciousness is called idealism.

> I believe that consciousness and its contents are all that exists. Space-time, matter and fields never were the fundamental denizens of the universe but have always been among the humbler contents of consciousness, dependent on it for their very being.
> Donald D. Hoffman, *professor of cognitive science*

---

* See page 25, 4.

† With the discovery of the DNA molecule, biology had indeed begun to turn itself into a testable science like physics, rather than a mostly descriptive one like botany.

All the choirs of heaven and furniture of the earth, in a word, all
those bodies which compose the mighty frame of the world, have
not any substance without a mind.
*George, Bishop Berkeley, philosopher*

I do not think that mind exists in the physical world, I rather think
that the physical world exists within mind.
*Jean-Markus Schwindt, theoretical physicist*

For after all, my mind had to be a single thing; or perhaps there is
only a single mind, in which everybody has a share, a mind to
which all of us look, isolated though each of us is within a private
body, just as at the theatre, where, though every spectator sits in a
separate place, there is only one stage.
*Marcel Proust (1871–1922), Remembrance of Things Past*

22 | The plasma physicist Robert G. Jahn believes that the interconnec-
tion between consciousness and the environment is so intimate that in
effect we create the reality we live in. It is the ego that creates the
illusion of separation between things.

23 | The physicist Erwin Schrödinger, who was influenced by Vedanta
and Schopenhauer, believed that there is a single consciousness that we
are each a part of. That feeling you have that you are a thing apart is a
delusion Hindus call *maya*. What we see as separate consciousnesses
are different aspects of the one thing. Schrödinger believed that the 'I'
persists even if all memory fails, and that the laws of nature are
contained in the 'I', not the other way about. He found 'distasteful' the
idea that consciousness dies with the body, but rejected the idea of
multiple separate and immortal souls. In physics the 1920s were the
1960s.

24 | Reality is best understood subjectively, said the philosopher Henri
Bergson. Reality is imprecise, said the physicist Werner Heisenberg.

Mind is woven into the fabric of the universe.
*Freeman Dyson, physicist and mathematician*

The mind is its own place, and in itself
Can make a heav'n of hell, a hell of heav'n.
*John Milton (1608–74), Paradise Lost*

25 | Carl Jung thought matter and psyche might turn out to be the same thing, one seen from within and one from without.

26 | Jean-Paul Sartre said that consciousness is nothing, because only objects of consciousness have existence.

27 | It may turn out that trying to understand consciousness by looking at the workings of the brain is like trying to find an explanation of why you feel as you do when you listen to Beethoven's *Hammerklavier* sonata by looking at the insides of a piano. You do not need to know how a piano works to play a piano. You do not need to know how the piano works, nor how to play a piano, in order to listen to a piano sonata.

28 | The philosopher Colin McGinn believes that consciousness will forever evade our understanding because of the constraints that the structure of the brain imposes on our ability to solve certain problems, but that hasn't stopped scientists from coming up with theories of how the universe was created, even though our ability to describe the universe is presumably limited by our being in it.

29 | No one knows when the advances will come in the understanding of consciousness, whether it will be in less than a hundred years or more than a thousand years. It may be that we will eventually come to understand why we cannot understand. And that in itself is a kind of understanding.

# SECTION 4

# On the self

The mind, by which I am what I am, is entirely distinct from the body, and even ... easier to know than the body, and moreover, that even if the body were not, it would not cease to be all that it is.

*René Descartes (1596–1650), philosopher*

1 | When Descartes tried to find a firm foundation to the world, he found one certainty on which to build his philosophy: that there is a self that exists. In a Cartesian world I am certain of my own consciousness, but uncertain of yours. There appears to be no way of reaching agreement that we both exist using objective scientific testing.

2 | Descartes – he was, after all, taught by Jesuits – agreed with Thomas Aquinas that the human soul is immortal and exists independently of the body. Descartes separated out thinking from material being, and so put consciousness out of bounds for three hundred years. He divided the world into the machine of Nature (not just the machine of Nature's physical processes, but the machine-like nature of all living creatures, excepting man) and the divine nature of man. Science took up the body, rejected the mind, and gave the soul to religion.

Science was concerned with the objective realm of facts; religion and the arts with the subjective realm of values, aesthetics, morality and belief. Science got the better part of the bargain, since it got practically everything, as defined in its own terms.

*Rupert Sheldrake,\* biochemist*

3 | A thought experiment in neuroscience: If we could make a perfect copy of a human being, would self-awareness naturally arise out of the complexity of the neural machinery of the brain, or is it possible that our doll might be physically identical but without self-awareness? Most neuroscientists deny the possibility of zombies. If we could make an exact material copy of you, that copy would have a sense of self, just as you do.

If there were such machines which had the organs and appearance of a monkey or some other irrational animal we would have no means of recognising that they were not of exactly the same nature as these animals.

*Descartes*

4 | Descartes used a form of the zombie argument in order to separate animals from humans. Only humans have souls. Without souls, animals are no more than automata. These days, minds and souls have become conflated into a single mystery. Today the debate is about whether animals have minds or not.

5 | Descartes came to the conclusion that the mind can be trusted; Freud that it can hardly ever be. Freud undermined the efficacy of introspection. It is possible to investigate the motives of other people, but we are invariably misled when we attempt to introspect on our own motives, was his point. Much of our conscious self is remarkably

---

\* Best known for his research into parapsychology.

private. We hide ourselves away from others, but continually give ourselves away, which is where psychology and psychiatry come in.

6 | If there is a persisting self, where is it, and how does it exist independent of physical reality? Descartes said it was clear that mindstuff existed separately from bodystuff, but was unable to describe what mindstuff is, nor what the mechanism is through which the two come together in a human being. He failed to bridge the dualist gap. He thought the mind was somehow housed in the pineal gland. He was wrong.

7 | Francis Crick discovered the DNA molecule, and so effectively answered the question, what separates the living from the non-living? For the latter part of his life, he attempted to answer the question, what separates the conscious from the unconscious? He continued the Cartesian search for a physical substrate, focusing on an area of the brain called the claustrum, a sheet of brain tissue beneath the cerebral cortex, as the place where experience is brought together. He believed that the claustrum behaves like the conductor of an orchestra.

> I would like to be a straightforward identity theorist – that somehow the experience just *is* the brain activity; but I cannot for the life of me see how it could possibly be.
> Susan Blackmore, psychologist

8 | Whatever we mean by consciousness is intimately bound up with whatever we mean by the self. The self has a sense that things matter, whereas to matter nothing matters.

9 | The inside experience of not being here, not embodied, neither corporeal nor real.

10 | Our conscious experience of the world is subjective, and we haven't known how to do subjectivity in science. If we are to explain

consciousness it seems as if we will have to work out how to make subjectivity objective, or modify science to include subjective tools. Either way, this heralds a revolution in biology and a new way of doing science. Perhaps not surprisingly, materialists do not have much time for such an approach.

> Nobody … would think of submitting a paper that said, 'Well, I introspected under the following conditions and these are the things I thought.'
> *Daniel Dennett*

> [The] scientist of tomorrow will pay special attention to his own inner life, subjecting it to analysis with a precision instrument created out of himself.
> *Fernando Pessoa (1888–1935), The Book of Disquiet*

11 | Introspection was the tool Descartes used to get at the truth, arguably with disastrous results.

12 | The self cannot be unravelled, it seems. We are at best a parliament of selves somehow brought together into the sensation of a self. There is nothing that answers to the concept of self, and so the self cannot exist. We are lonely ghosts haunting a material world.

> Each of us is several, is many, a profusion of selves. So that the self who disdains his surroundings is not the same self who suffers or takes joy in them. In the vast colony of our being there are many species of people who think and feel in different ways.
> *Fernando Pessoa, The Book of Disquiet*

13 | There is a voice inside you that you just can't fool. Is that how we know we are sane?

Know thyself.
*Inscription at one end of the Temple of Apollo at Delphi**

To thine own self be true.
*Polonius, in Hamlet*

14 | *Just* be yourself. Was there ever a harder piece of advice to follow?

15 | Jill Bolte Taylor is a neuroscientist who has spent many years of her life cutting up brains trying to find where we might be located in a material world. On the morning of 10 December 1996 she was on her exercise bike when something odd began to happen. At first she couldn't work out what was going on, and though she felt dazed and disoriented she tried to carry on as normal, getting into the shower, balancing her weight against the shower wall, hoping the faintness would pass. But whereas normally in a dazed state the world becomes confused or muted, somehow that morning the world was more clamorous than usual. The water surely sounded louder, like a roar almost, and more curious still, she began to feel that she was becoming aware of the inner activities of her own brain and body. It was as if she could hear the right-hand side of her brain talking – or chattering, as she described it – to the muscles in her limbs as she tried to keep herself upright. As reality continued to rearrange itself about her, the more usually conscious part of her brain began to realise that she was having a stroke. A massive stroke, as it turned out. She was thirty-seven, and it would take eight years for her to make a full recovery.

Even during the stroke she realised that she was in a unique position: a brain specialist witnessing her own stroke from the inside. She knew that the blood flooding into the left and logical side of her brain was destroying the neurons it came into contact with. Blood normally flows only on the surface of the brain, where the grey matter is. The

---

* 'Nothing in Excess' was inscribed at the other end.

white interior of the brain – where all the wiring is housed that connects each neuron to up to a thousand others – is blood-free. The stroke was destroying the hardwiring of her life's experience. Bizarrely, as the left side of her brain began to close down, she sensed herself moving into what she describes as the 'euphoric' right side of the brain. She felt herself turn into something immaterial that was spreading outside the confines of her skin until 'It was obvious to me that I would never be able to squeeze the enormousness of my spirit back into this tiny cellular matrix.' She escaped from the confines of her body like a genie from an uncorked bottle.

Jill Bolte Taylor's account of her experience of stroke and recovery is uniquely valuable as evidence from a specialist who can relate the objective physical experience of what was happening to her to a subjective account of what it felt like to be having that experience.

One thought fills immensity.
*William Blake*

16 | Science has generally been mistrustful of introspection and personal experience. Such mistrust denies us our deepest human experience of the world, rather as post-structuralism denies authors the validity of their own works. Science looks for collective knowledge, just as post-structuralism shifted the perspective from the genius of individual authors to the collective authority and context of culture and society. But science has a remarkable ability to reinvent itself, and is, in any case, not one single discipline, but many approaches. We may be witnessing a shift towards holistic methodologies to be put alongside reductive ones.

The systematic denial on science's part of personality as a condition of events, this rigorous belief that in its own essential and innermost nature our world is a strictly impersonal world, may, conceivably, as the whirligig of time goes round, prove to be the very defect that our descendants will be most surprised at in our

own boasted science, the omission that to their eyes will most tend to make it look perspectiveless and short.
*William James*

We remain unknown to ourselves, we seekers after knowledge, even to ourselves: and with good reason. We have never sought after ourselves – so how should we one day find ourselves? It has rightly been said that: 'Where your treasure is, there will your heart be also,' our treasure is to be found in the beehives of knowledge. As spiritual bees from birth, this is our eternal destination, our hearts are set on one thing only – 'bringing something home'. Whatever else life has to offer, so-called 'experiences' – who among us is serious enough for them? Or has enough time for them? In such matters, we were, I fear, never properly 'abreast of things': our heart is just not in it – nor, if it comes to it, are our ears! Imagine someone who, when woken suddenly from divine distraction and self-absorption by the twelve loud strokes of the noon bell, asks himself: 'What time is it?' In much the same way, we rub our ears after the fact and ask in complete surprise and embarrassment: 'What was that we just experienced?', or even 'Who are we really?' Then we count back over in retrospect, as I said, every one of the twelve trembling strokes of our experience, our life, our being – and alas! lose our count in the process … And so we necessarily remain a mystery to ourselves, we fail to understand ourselves, we are bound to mistake ourselves. Our eternal sentence reads: 'Everyone is furthest from himself' – of ourselves, we have no knowledge …
*Nietzsche, On the Genealogy of Morals*

17 | The psychologist and writer on neuroscience Susan Blackmore has been meditating for years and reflecting on the nature of her self. She has reached the stark conclusion that there is nobody there. She has come to accept that every time she thinks she exists, 'this is just a temporary fiction and not the same "me" who seemed to exist a

moment before, or last week, or last year. This is tough, but I think it gets easier with practice.' She says she is now so convinced that there is no self there that she catches herself thinking in the third person. In a restaurant she asks herself: 'I wonder what she's going to choose,' though presumably she also accepts that neither is there anyone doing the wondering, nor anyone wondering who is wondering about who is wondering about the person ordering her food ...

18 | Susan Blackmore's conclusions and methodology may be unorthodox scientifically, but her techniques and conclusions are similar to those Buddhists have known of, and reached, for centuries. Zen Buddhists use such ways of thinking to attain enlightenment. The Buddha teaches that the self is an illusion: not that it doesn't exist, but that it is not what we think it is. Actions exist, says the Buddha, and also their consequences, but the person that acts does not. The self appears to exist, says Blackmore, because the brain continually makes up stories that convince us that there is an 'I', special and separate from the rest of our busy body. Not that the problem goes away even if we throw away the idea of a self: 'I think there can't be two separate things, the room we're sitting in, and my experience of the room; somehow the two have to be integrated, and I don't know how ...'

If, out of such philosophical musings, the existence of the self is left undecided, the possibility of free will appears to be easier to determine. There isn't any. Once we understand that all the actions of our body 'are the consequences of prior events acting on a complex system; then the feeling of making free conscious decisions simply melts away'. *Simply!*

Soon I became aware, to my surprise, that every thought I conceived was suggested by an external impression. Not only this but all my actions were prompted in a similar way. In the course of time it became perfectly evident to me that I was merely an automaton endowed with the power of thinking and acting accordingly.
*Nikola Tesla (1856–1943), engineer and physicist*

# Meat → mind

## A: Humans as machines

We are conscious automata.
*Thomas H. Huxley (1825–95), biologist*

Just a bunch of neurons ... You, your joys, and your sorrows, your memories and your ambitions, your sense of personal identity and free will, are in fact no more than the behaviour of a vast assembly of nerve cells and their associated molecules.
*Francis Crick*

Survival machines – robot vehicles blindly programmed to preserve the selfish molecules known as genes.
*Richard Dawkins*

We have been created by the process of evolution, both genetic and cultural. And what we're now trying to do is to reverse engineer ourselves, to understand what kind of machine we are that this is true of.
*Daniel Dennett*

The cybernetic structure of a person has been refined by a very large, very long, and very deep encounter with physical reality.
*Jaron Lanier*

[Sentience is] an emergent characteristic of a certain variety of information processing.
*Brian Greene*

Ever since I've been a child, I've wanted to be a robot. I think one of the great difficulties of human life is that one's life is inhabited by uncontrollable desires and that if one could only be master of those and become more like a robot one would be much better off.
*Kevin O'Regan, experimental psychologist*

1 | If we wish to reverse engineer the human machine and shortcut the long path evolution took through deep time, we will need to find out what kind of machine a human being is.

2 | Every generation has described humans by analogy to the best technology available at the time. For the physiologist Charles Sherrington (1857–1952) the brain was 'an enchanted loom where millions of flashing shuttles weave a dissolving pattern, always a meaningful pattern though never an abiding one; a shifting harmony of subpatterns'. In our own age we compare brains to computers,* not looms. We attempt to turn humans into zeroes and ones – neural circuits processing as if they could be reduced to the electronic circuits of a computer – just as in the past we have attempted, with varying degrees of explanatory power, to reduce humans to clocks or steam engines. Future generations will no doubt appeal to technological advances we cannot predict or imagine from this vantage point. The modern idea that the brain is computational is an assumption, or a belief, that reflects our best current technologies.

---

* The universe, too, is seen as some kind of computer, with the laws of nature as its software. Universes are simulated on computers, in a sort of hit-or-miss process, by theoretical physicists and mathematicians hoping to alight on one that looks like this one.

3 | We humans have within us, Tristram Shandy's father said, using the language of the clockmaker, some 'great and elastic power … of counterbalancing evil, which like a secret spring in a well-ordered machine, though it can't prevent the shock – at least it imposes upon our sense of it'.

4 | The mathematician Paul Erdös (1913–96) once described a mathematician as 'a machine for turning coffee into theorems', the intentional bathos (and pathos) of the remark serving to remind us that humans are more than machines even if mathematicians, according to Erdös, sometimes are not.

## B: The brain as a machine

1 | If we are to have any hope of ever understanding consciousness, we need first to understand how the brain works. A reductive account of how flesh could have been woven into mind is a prospect still some way off into the future, but there is much to encourage those who believe that such a story can be told.

2 | We do not know how to explain the dualism of consciousness and unconsciousness, and there is no general agreement even on how to define consciousness, but such philosophical concerns do not stand in the way of doing science. It is far too soon to give up. Consciousness may turn out to be like aether or vitalism.* Out of a deeper understanding of nature, the concept may simply become irrelevant.

---

* Aether was the name given to the medium through which light travels. What aether could be made from was a seemingly insoluble problem for centuries. Then Einstein posited that light does not move relative to anything, and that must include relative to aether too. In effect, his special theory of relativity ruled aether out of existence. How light can be a wave but not a wave *in* anything became a problem for philosophers: physicists moved on to address other problems about the nature of light. Vitalism, as has already been noted, effectively disappeared with the discovery of DNA.

3 | Neuroscience generally attempts to reduce consciousness to physical processes happening in the brain. It may be that a full explanation will also require explanations at larger scales, but in the brain is a good place to begin to look for answers.

4 | Biological descriptions of the brain could easily become overwhelmed by detail, with no way of deciding what is important and what is not. At one time it was thought that by recording what every neuron in the brain is doing, understanding of how the brain works would somehow follow; instead, what we would end up with is an intractable computational problem. Effectively we would end up with another brain. In order to begin to understand the brain we have to understand what fundamental principles govern and determine its activity. The brain has to be somehow modelled if we want to make predictions about it that we can then test. We have to reduce the brain, but reduce it too much and you end up with a computational problem, don't reduce it enough and you can't make any predictions.* If there were not reductive principles, science would be impossible.

The brain represents problems to be solved at many levels, from molecules to cells, cells to circuits, circuits to systems, from systems to organisms, from organisms to societies. But no one yet knows how these levels relate, and that is a profound problem. In this, the brain – the most complex structure we know of *in* the universe – is no different from that other complex structure, the universe itself. Physical theories are all very well as descriptions of the universe overall, but when we

---

* There are many combinatorial explosions of different kinds in science. It is very hard to predict from an amino-acid sequence how a protein will fold. There are twenty different kinds of amino acid. To make a protein, say, two hundred amino acids long there are $200^{20}$ possible proteins that need to be examined. For the sake of comparison, $200^{20}$ is close to $10^{260}$. There are 'only' $10^{80}$ elementary particles in the whole of the visible universe. Even if we could predict how a protein folds as it is made, then there is the problem of working out how proteins interact with other proteins. Perhaps not surprisingly, most medicines are fairly simple molecules compared to the proteins they are designed to interact with.

look at the universe too close-up or from too great a distance *life* gets overlooked.

> This neuron fires because of this channel opening. We're good at that kind of explanation, but putting it all together is harder. We don't know how far we'll get. We may have to throw up our hands and say, we don't know what the integrated parts mean.
> *From a conversation between a biologist and the author*

5 | We have hardly begun to understand how the brain works. It will surely be a long time before we get to the stage of throwing up our scientific hands (and it may never come to that). In the meantime there is much left to do, and much that has already been achieved.

6 | In the eighteenth century the Italian anatomist Luigi Galvani (1737–98) used electricity to make a dead frog kick its legs. His experiments established the understanding that the nervous system runs on electricity. The brain can be thought of as a machine that turns sensory input into electricity.

7 | The brain is constructed out of cells called neurons. A neuron totals up the inhibitory and excitatory input from a number of other neurons, and only fires if a certain threshold is passed. The firing is called the action potential. The potential passes along a thread called an axon that divides and becomes the possible input for another neuron; and, because the axons are long, possibly a neuron some distance away. The signal causes a neuron to release neurotransmitter molecules that then activate or inhibit cells in a circuit. We know that neurons are binary, that they either fire or do not fire, in all-or-nothing fashion. Whatever else the brain is, it is some kind of electrical system.

8 | i The human brain is the most complex structure in the universe that we are aware of, made of some one hundred billion neurons, each capable of making thousands of connections to other neurons. The

number of possible connections that can be made in a single human brain is much larger than the number of elementary particles in the visible universe.

ii In the human brain there are several thousand miles of axons.

9 | There are many different kinds of specialised neurons: motor neurons that make a muscle fibre contract, neurons that separately detect lines at certain angles, depth, distance, motion, and colour.

10 | It is assumed that the brain is a network made out of many separate circuits that are integrated into hierarchies of micro and macro circuits.

11 | The body is riddled with nerves. Overall, the nervous system is divided in two: the central nervous system that is housed in the brain and spinal cord, and the peripheral nervous system. The peripheral nervous system is itself divided: into the systems (collectively the autonomic nervous system) that work largely under the radar of consciousness, controlling unconscious activities like respiration, salivation, heart rate, perspiration, urination, sexual arousal and so on;* and the sympathetic nervous system that controls flight-or-fight responses and tries to maintain stability within whatever changing environments the body finds itself.

12 | The brain is activated by sensory input. The sensory organs detect limited amounts of the world at the outer reaches of the body. The information is codified† and transported to a number of inbuilt

---

* It is also possible to direct our conscious attention to many of these bodily functions.

† Colours are names for different wavelengths of light, and as such can be reduced to numbers. Sound is interpreted out of pulses of air that can also be encoded mathematically. Images and sounds are baroque constructions made by the brain.

structures in the brain that make meaning out of these samples of the world. What we make of the world is an abstraction, not a replication. The world is constructed out of layers of neural signals in hierarchies of firing patterns. These layers are also mediated by layers of chemical signals that produce physical change in the circuits. The neural signals translated out of the world's input are amplified and damped in all kinds of ways by our nervous system.

13 | Nerve cells on the skin and in the sense organs – called sensory neurons – respond to stimuli from the outside world: to pressure, light, sound waves, and various chemicals. Motor neurons are located in the brain stem and spinal cord. Their axons reach out of these regions to control the activity of muscle and gland cells. Sensory information received at the skin travels along nerves to the spinal cord and up to the spinal stem, and from there into the brain, where it is turned into motor commands that are relayed back to the muscles via other nerve pathways. Reflexes are processed in the spinal cord and lower parts of the brain. The most common type of neuron in the brain is the interneuron; collectively they mediate activity between sensory and motor neurons. Perched on the brain stem are the hypothalamus, the thalamus and the heavy outer layers of the cerebral cortex. The hypothalamus is associated with emotion and the secretion of hormones, also body temperature, hunger and thirst, sleep and circadian rhythms. Buried inside the cerebral cortex are the basal ganglia, the hippocampus and the amygdala. Basal ganglia regulate motor performance. The hippocampus consolidates short- into long-term memory. The amygdala controls emotional reactions, and memories of emotions. The cerebral cortex controls higher mental functions like perception, language, planning and awareness. It is where conscious decisions are activated, and where long-term memory is stored after it has been consolidated in the hippocampus. The cerebral cortex is the covering of the mammalian brain. In large mammals it is folded. In humans it is heavily folded. The brain processes raw sensory input, which, passed through various structures of the brain collectively

known as the limbic system (which includes the hippocampus and amygdalae), is turned into emotions – most crudely pain or pleasure. In humans some of that emotional information may become further processed, particularly in the pre-cerebral cortex, the most recently evolved part of the brain. Intellect, we now know, rests on a bed of emotion. Humans are feeling organisms that also think.

14 | The brain changes all the time, whether we like it or not. Fats and proteins change as soon as they are made. Molecules around the synapses are changed for other molecules every hour or so. Human bodies change cell by cell, like the exterior of some cathedral replaced stone by stone over the ages. The cerebral cortex, however, is as old as you are. Except for a small number of a particular type of neuron, no new neurons are made during adult life.

Except for the retina, the brain consumes more energy relative to its mass than any other part of the body.

The primary task of the brain is to help maintain the whole body in an optimal state relative to the environment.
*Rita Carter, The Brain Book*

15 | Touch is felt by skin sensors that translate the energy of the touch into an electrical signal that passes along a precise route to the brain; from there it goes through various relay stations in the brain stem and thalamus before ending up in the somatosensory cortex. If you are touched in adjacent places on the body, say, on one finger and then the next finger, the signal travels along adjacent nerve pathways,* with the result that there is a scaled-down one-to-one map of the body in the brain. A mild touch will generate two or three action potentials per second, whereas a pinch or a knock might cause a

---

* Many sensory, motor and cognitive functions are processed along more than one neural pathway. The information is processed in parallel across the brain and somehow integrated.

hundred neurons to fire. In a violinist's brain the hands are more prominently represented in the cerebral cortex than the norm. For a right-handed violinist, the mapping of the left hand that deals with fingering is even larger than that of the bowing hand. How the body is represented in the cortex depends on what parts of it are used, and with what intensity and complexity. Conversely, imagination alone can improve the strength of body parts. Merely imagining that you are moving your finger is enough to increase the finger's strength. Doing this for just ten minutes a day for four weeks increases the finger's strength significantly.*

> Let it suffice to affirm, that of all the senses, the eye ... has the quickest commerce with the soul – gives a smarter stroke, and leaves something more inexpressible upon the fancy, than words can either convey – or sometimes get rid of.
> *Laurence Sterne, Tristram Shandy*

16 | Light is made out of particles of energy called photons. When light strikes the retina, energy is transferred from the photons and causes changes of voltage at the cellular level. This energetic information is sent along parallel neural pathways to the visual cortex where the information is processed. The visual cortex constructs a 3D coloured-in world out of these signals. Curiously, the visual cortex is at the back of the cerebral cortex, the furthest part of the brain from the eyes.

17 | Visual information is sent from the retina along two main pathways, fast and slow. The fast track instigates responses to the world ahead of our conscious apprehension of it. The slow path leads from the primary visual cortex to the temporal cortex and builds up perceptions of the world. This takes time. The energetic information is scattered

---

* Perhaps I can imagine going to the gym, and save both time and money.

across the visual cortex at the back of brain in at least thirty different maps.*

18 | We can lose the use of single components of our vision machinery. For people with depth agnosia, for example, all objects look like cardboard cut-outs. Those with so-called blind sight have lost the ability to process the world in the slow way. They unconsciously respond to the world as if they are sighted, but claim consciously in words that they are blind.

19 | Smell is different from the other senses. Smells are apprehended directly. Smell steals a part of the thing itself. Smell an apple and you absorb a tiny part of the apple. Smell requires direct contact with the thing itself. Perhaps this is why we find smells so repellent or alluring: there is no mediation between the thing and ourselves.

The brain contains thousands of neurons dedicated to smell. Each type is capable of detecting a particular kind of smell, and contains idiosyncratic cup-shaped proteins that are shaped in just the right way to catch a particularly-shaped molecule. It is the capture of those kinds of molecules and no others, and the resulting neural signal that is triggered, that represents what we call the particular smell of something. Where we don't have the right-shaped cups, those substances have no smell.† Smells can be combined, but only out of a fixed number of possibilities. Smells don't grade into each other as colours do. To detect a smell is analogous to the brain looking up the word associated with that particular incoming molecule in a dictionary. If

---

* In the 1950s Stephen Kuffler discovered that there are certain retinal cells that measure the contrast between light and dark, not absolute levels of light. All these aspects of an object are processed separately and then integrated in the higher regions of the cortex into a single unified experience. How a single unified experience is achieved is not known. Some would say, nor is it known what a single unified experience could mean.

† Some air-fresheners work by interfering with the nose's ability to smell. The smell doesn't go away, only the ability to smell it.

there is no word there, there is no smell. The smell mechanism was discovered by the neurobiologist Cori Bargmann (b.1961), first in nematode worms. That the structure is largely preserved in humans tells us that it has its origin in deep time. The nematode C elegans is made out of just 959 cells, 302 of which are neurons. In order to identify which of the neurons controlled smell, Bargmann had to destroy each neuron, one at a time, and record whether or not the worm's sense of smell was affected. Seeing that all the neurons look the same, this is quite a task. As the head of her lab, H. Robert Horvitz, put it: 'It's like knowing each grape in a bunch is different, but not quite being able to see it.' As so often happens in science, only by looking closely at what looks indistinguishable can difference be identified. By accident Bargmann also identified the neuron that controls round-worm hibernation. Horvitz once told her that she could think like a worm. The world of C elegans is dominated by smell. It has 2,000 odour receptors, twice as many as in mice. Bargmann discovered that each receptor corresponds to a single smell. This was a major break-through that showed how smell works in all animals. Soon afterwards Richard Axel and Linda Buck showed that in mice olfactory nerves have studs on their ends that detect specific odours. Dogs have more smell receptors than humans, twenty to forty times as many, but ours are more connected. We can train our sense of smell. Practice is the key to many human skills.

20 | There are five types of human taste receptor cells. They differenti-ate between salt, sweet, sour, bitter, umami.* Phenylthiocarbamide is a chemical that has the unusual property of tasting very bitter to around

---

* Umami was accepted as a separate kind of taste in 1985. It is the taste that gives meat stock its particular mouthwatering quality. In the late nineteenth century, the French chef Auguste Escoffier revolutionised cooking by combining foods of different tastes. His slow-cooked veal stock, the base of much of his cooking, was umami. Breast milk is as umami as meat stock.

three-quarters of the world's population,* and virtually tasteless to the rest. So does it have a taste or not?

21 | Visual light is a small part of the electromagnetic spectrum. If we saw a wider range of radiation we might see radio waves rather than hear them, or see infra-red rather than feel it as heat. We would apprehend the world differently, as many animals do, and as, presumably, aliens do. There are animals that 'see' disturbances of the air. Bats navigate by interpreting the wavelengths of sound. By emitting high-frequency (i.e. short-wavelength) sounds, bats are able to detect small objects. Anything much smaller than the wavelength of that sound cannot be seen. The wavelength of sunlight puts a limit to what humans can detect with the naked eye.

22 | Moonlight is not energetic enough to show the world in colour, but we do not take the world to be any less real on moonlit nights. Imagine if we had only ever seen the world by moonlight, and there came a day when a human being made an artificial light powerful enough to transform the world into colour. Humans might be tempted to suppose that it was they who had transformed the world, rather than that the world had possessed this quality all along. Moonlight is a kind of reduction, but seeing the world in colour is not its completion.

23 | We do not store a rich representation of the world from one moment to another. The world is thrown away every time we blink. From a reductive, evolutionary, point of view the brain evolved first as a survival machine. If it is to survive in a world that suddenly changes, an animal must be alert to those changes that are of possible harm to it. We evolved to hone our attention; not to take in the whole scene but to notice key elements, and what changes. We take in what is important and jettison the rest.

---

* The percentage rises to 98 per cent among the indigenous peoples of the Americas.

24 | Humans create a sensation of movement by having their eyes move continually about the scene. (Frogs, conversely, do not see anything that does not move.) Each movement is called a saccade.

25 | We do not appear to have any conscious experience of the world unless we direct our attention to it. Consciousness is like the light in the fridge. It seems always to be on only because it comes on every time we open the door, though in fact mostly the light is off. And yet the world constantly jumps back the moment we try to see where it has gone to: a world consistent with how it looked when we last looked. When we do not attend to the world it is hard to say whether we see anything at all. People who are good at meditating can empty out their minds, but a sudden noise or disturbance is often enough to bring the world flooding back in.

26 | Even when we do direct our attention to something in particular we can fail to notice key elements. A now famous experiment asks participants to watch a film of basketball players, and to count the number of times the ball is thrown between the players. In the background a man in a bear costume dances across the screen. Fifty per cent of those who watch the film are so distracted by the task of counting that they fail to notice the bear.

27 | Our attention can also be surprisingly acute and persistent. We tend to do better than we think we will in an experiment that asks us to direct our attention to a tray of objects. When asked later to recall what was on the tray, we typically perform rather poorly. Being unable to name many objects directly, we may be tempted to come to the conclusion that we have taken in very little. But if we are then shown a large array of objects and asked to indicate which were and which weren't on the tray, we invariably perform that task very well, which can only mean that somewhere we do know what was on the tray, even if we cannot get to the information very easily without the reminder of the object itself. An elaborate version of the tray game has been conducted

under laboratory conditions. The test takes a week to complete. Ten thousand images are presented to a subject, one after another. The subjects are generally not very good at recalling what they have seen unprompted, but are very good at deciding between possibilities of what might or might not have been seen before. A more recent version of the same test using 2,500 objects shows that subjects can say with a 90 per cent success rate whether or not a subsequently presented object has been shown before. The subjects even performed well years later, being capable of such small distinctions as deciding between a bell with a narrow handle and one with a wide handle.

# SECTION 6

# On perception

The fool sees not the same tree the wise man sees.
*William Blake*

Who ever *felt* a *single* sensation? Is not every one at the same
moment conscious that there co-exist a thousand others in a
darker shade, or less light ...?
*Samuel Taylor Coleridge (1772–1834), poet, in Notebook 21*

1 | You look at a tree and specialised neurons of different kinds fire in
response to the tree's colour and depth, to the edges of its branches and
so on. There is no single group of cells that responds to the tree overall,
and yet we perceive the tree as something cohesive. This is the binding
problem. We do not know how it happens, nor even how it could
happen. What could it mean from the brain's perspective to respond to
the tree overall?

2 | The brain models space out of various sensory inputs in different
ways and in different contexts. So how do we have a consistent idea of
space? The binding problem. How does the distributive system of the
brain make the apparent unity of conscious perception?

Of all the gifts bestowed upon us, colour is the holiest, the most
divine, the most solemn.

*John Ruskin (1819–1900), art critic*

3 | To be human is to be private: from each other, from nature, and
from ourselves. Our consciousness creates the private and subjective
world that we call self. Philosophers say that that private world of how
things appear to be is constructed out of ineffable attributes of the
world called qualia. Our emotional responses to colour, pain, taste, or
music are examples of such attributes of the world. Qualia (singular
quale) cannot be accounted for by physics, it seems. It remains an open
question whether they can be accounted for by physiology.

4 | When you look at a red rose you respond to the colour red. By
comparing the firing of the same kind of nerve cells in the visual
systems of different human beings, it is possible to show that we all see
the same colour red. But what we cannot compare are our emotional
feelings in response to the red of the rose. That emotional response is
an example of a quale. Some materialists say that the word qualia gives
form to things that do not exist and do not need accounting for.

As individuals we can make a cultural connection with other individual human beings by talking about our response, or writing a poem
about it, but it is hard to know what a scientific test would look like. We
do not know whether or not the cells that fire in response to the edges
of a rose, or to its colour and depth, have anything to do with the
emotions we also feel.

What we understand by the experience of seeing (feeling) a rose is
calculated by our brain. In order to understand how this happens we
have to understand how the brain works and what we mean by experience. We need to study inner and outer worlds, a first for science.

5 | Photographers and painters know that what we see is not the thing
itself but the light falling on the thing.

6 | i The various neural responses to looking at a tree – edge, colour, depth, etc. – are called neural correlates. Neurons fire in the brain in response to observing the world. We now have machines that can record these neural responses. EEG (electroencephalography) records the electrical activity of the cortex, PET (positron emission topography) tags oxygen and glucose in the brain with radioactivity, and MEG (magneto-encephalography) measures the electrical activity of brain cells. MRI (magnetic resonance imaging) measures changes of blood flow in the brain. When traced over time it is called functional MRI, or fMRI. FMRI shows the brain in new detail, seen from different angles and in slices that can be built up to give a picture of overall brain activity through time. The time lag also makes it a crude measurement. Some neuroscientists think that such scans show mental events as brain activity; others point out that they show only correlates, not necessarily anything causal.

ii FMRI measures changes of blood flow in tens of cubic millimetres of brain tissue. There are around 100 million synaptic connections in such a volume. The scans make psychological studies of the brain more scientific, but in a limited way.

iii FMRI records localised brain activity and ignores the distributive function of the brain. Most of the evidence we currently have about how the brain works has been gathered from the study of brains that are damaged in some way. Damage to particular regions tells us that those regions must be necessary to whatever behaviour has been impaired. Such lesion evidence shows us that there is behaviour that is localised, and not distributed over the whole brain. It is hypothesised that this local activity is underpinned by local neural circuitry. Knowing that there are localised parts of the brain required for certain functions is a big advance from not knowing that. Nevertheless, most brain activity is not local. Even the association of the words 'chair' and 'sits' involves many regions of the brain. The simplest thought involves highly distributed brain activity. When it comes to recording how the brain responds to, say, making a moral choice, we get lost in complexity. Many different areas of the brain are implicated, some a long way away from the first circuits that fire. Science looks for reductive explanations, but humans continually thwart scientific investigation by losing themselves in complexity.

iv FMRI measures additional activity. Large parts of the brain are already lit up. Even repeated measurements of the same simple human activity have given highly various results. Repeated measuring of finger tapping, for example, showed sometimes zero correlation between the brain activity recorded in different scans. That we can see close-up images of the brain responding to the world is, of course, extraordinary. And the possibility fMRI offers of communicating with patients suffering from locked-in syndrome comes close to whatever we might mean by miraculous.

7 | Scientists have in a very few years found many neural correlates to action and to thinking. We can see what parts of the brain light up when certain stimuli are applied to the body or to the brain, or when we are instructed to have certain thoughts. It is easy to think that we are well on the road to working out how mind arises, and to forget that we have no causal mechanism that relates those correlates to actual thoughts. Not only that, but we still have no idea how there could be such a mechanism.

8 | Rare rhythmic neural firing patterns have been seen in Tibetan monks and London taxi drivers, but the cause isn't in the neurons: the firing patterns have become ingrained in the circuitry out of a particular life lived.

9 | Our lives are recorded in the brain, and there are machines that can see that record being laid down, but that does not prove that our experience as humans is only neural. Being human must surely also exist at larger scales: body, bodies. The brain's firing patterns are the result of the interaction of brain and body, and environment.

Trying to discover the contents of our ordinary Wednesdays in the tropisms of the evolved organism as reflected in brain activity is like applying one's ear to a seed and expecting to hear the rustling of the woods in a breeze.
*Raymond Tallis*

10 | Materialists make strong claims for the future reductive understanding of cognition, and they may well turn out to be justified, but at the scale of cognition a reductive neural understanding of cognition still has a long way to go.

The future of the humanities[:] connecting aesthetics and criticism to an understanding of human nature drawn from the cognitive and biological sciences.
*Steven Pinker, cognitive scientist*

11 | Neuroscience has invaded all areas of intellectual investigation. There is even neuro lit crit. Investigators have looked at what parts of the brain light up when we read particular novels. What would be impressive is if from this information they could determine what novel is being read. We know that Henry James is a different kind of writer from Barbara Cartland, so it is hardly surprising to find out that different books cause different parts of the brain to light up in fMRI scans. The question is, what do the firing patterns mean and what can we predict from them? Could we one day isolate and artificially fire a set of neural circuits that gives the reader the sensation that she is reading Henry James when in reality she is sitting in an armchair staring into space? That day is a long way off, and there are reasons to suppose that we will never achieve this dubious ability. Who is to say, anyway, what Barbara Cartland is in relation to Henry James? That's what culture is: the collective agreement, and tolerance (or intolerance), of different opinions among different groups of human beings.

The apparent fact that the same brain areas are activated when we listen to pleasurable music and during sex confirms how uninformative imaging is. Techniques that cannot distinguish between hearing an organ played and having one's organ played with tell us little about them.

*Raymond Tallis*

12 | In a recent test a subject is asked to look at a line oriented at some angle unknown to the investigator. Using fMRI techniques to record the activity of specific neurons, the investigator can predict with 85 per cent accuracy the value of that angle. This is an extraordinary achievement, but it is very different from predicting whether the subject has uttered the word 'love', or indeed, is in love. Isolating human experience in a laboratory is as hard to achieve as the isolation of a quantum object made out of more than a few molecules. To be complex is to be hidden. Humans are the most complex structures we know of in the visible universe.

13 | It looks as if perception is a dynamic bodily activity that cannot be broken down into parts. Rather than being separate from what we experience, the physical world itself becomes part of the experience. The world exists, but what it exists as depends on how it is viewed. The world as we are conscious of it is a matter of perspective. We can change our perspective by changing how we attend to it. We might look decidedly for high windows, or look more intently than usual at trees, or at things of the colour blue. Some days we are angry, some days in love. The world looks differently according to lens and mood.

# SECTION 7

# On free will

1 | A famous experiment in neuroscience, first conducted in the 1970s, called the Libet test, seems to have confirmed beyond all reasonable doubt that free will does not exist. The test is named after its inventor, the physiologist Benjamin Libet (1916–2007).

2 | The subject, wired up to an EEG, is asked to move a hand at any moment of her choosing. The EEG measures the activity of the brain that arises when the brain is ready to make the move. This activity is recorded in an area of the cerebral cortex associated with voluntary movement, and typically shows that the brain is preparing to make the movement of the hand about half a second before the hand actually moves. This is as we would expect it to be. It takes time for the brain to process and translate the thought of making a movement into the actual movement. Neural activity has to move from the brain down the relevant neural pathway to the muscles that control the movement of the hand. A second measurement records the moment the subject acknowledges that she was first aware of the intention to move her hand. Shockingly, the first measurement – indicating the brain's readiness to move the hand – comes about a third of a second *before* the subject's conscious intention to move her hand. Subsequent refinements of the experiment have shown the time lag to be even greater. There seems to be no alternative but to accept that the brain has already decided how the body will respond before any conscious decision has been made.

3 | The Libet test shows that brain activity begins some significant fraction of a second before the move is willed. Whatever the brain selects to present to our conscious self is done later. Free will is ruled out. The body has already made a decision before we consciously decide what that decision is.

4 | Libet has suggested that rather than having free will, what we have is free won't. We have the choice not to act on an array of possibilities that are presented to us. It is free won't, and not free will, that directs our actions. The process is initiated unconsciously, but our conscious mind has the option of aborting the action. Which raises the question, why does the action need rubber-stamping?

5 | The Libet test forces us to the strange conclusion that we do not have free will, but only if we accept that there is a conscious self that lags behind a more attentive and unconscious body; that is, we are forced to accept the dualism of a conscious self that is separate, in some mysterious way, from the rest of the world. Free will does not exist, but only if consciousness does.

6 | The Libet test shows us how consciousness might exist with no power to act on the brain. Consciousness hovers like a cloud about us, telling us stories after the fact, to justify what the world and the body have already worked out together. Consciousness hums the self into existence.

7 | Philosophical enquiry and introspection tell us that there can be no such thing as a self. Measurement and observation of brain activity confirm that the brain does not operate as if being directed from some HQ of the I. Most of our responses to the outside world result in neural activity widely distributed across the brain. The brain is said to be radically parallel. And yet the Libet test shows us how there could be a self after all, and at the cost of free will.

8 | What could it mean to act freely in the world? Nothing can be entirely free of the world. If there were some thing or some being that was entirely free, then through what agency could that thing or being act on the world and still be free of it?

9 | Free will cannot exist in a material world of moving parts, yet everything science has uncovered about complexity, from gas laws, chaos theory and quantum mechanics, seems to guarantee the illusion of free will.

10 | Philosophically, free will is impossible, and scientifically it is, at best, an illusion.

11 | The notion of free will is incompatible with both determinism and indeterminism. If everything happens for a physical reason, we are not free. Neither are we free if everything happens by chance. The desire to have free will is the desire to be free of the machine of nature.

12 | Scientific measurements of humans isolate the subject in a frozen moment that is almost never true of life as it moves through time. In these individual moments we see clearly that we do not have free will, but these moments are attached to other moments that all bleed into the world, and in the world at large complexity reigns. Free will is not 'a twitch before a twitch', but a set of social obligations. To be human is to balance a sense of autonomy and being a part of the world and society.

13 | Humans have freedom because true freedom is not about isolated actions that can be measured and predicted, it is about the flow of life that gets lost in chaos and complexity. Science isolates what it can, and makes a joined-up reality out of the pieces. It is a matter of belief how complete we think that picture is, or ever could be: near enough to make no difference, or always with almost everything missing.

# SECTION 8

# On behaviour

## A: On understanding human behaviour

1 | The mathematician Pierre-Simon Laplace (1749–1827) thought that humans might one day be able to predict their own behaviour. For Freud the brain was a black box,* but he believed that eventually the box would be opened, that ideas in psychology would be based in the understanding of some organic substrate.† Like biology, psychology was at first better at generating questions than answering them. Some would say psychology is still like this.

It was not clear if a science of human behaviour was even possible, and for the first behaviourists the brain was a black box in which they had no interest. They were only concerned with what could be measured as input and output. They ignored mental life and focused instead on observable behaviour.

> Why did millions of men set about killing each other, if it has been known since the world began that it is physically and morally bad?

---

\* In science, 'black box' is a term used for any kind of system whose internal workings are mysterious or hidden. By concentrating instead on the inputs and outputs of the system, often progress can still be made; indeed, progress that may allow the contents of the black box to be probed at some future time.

† When Freud was a young man he spent time dissecting eels, looking for their testes. Perhaps if he had been born later, or biology had been more advanced in his own time, he would have chosen neuroscience over psychology.

Because it was inevitably necessary that in fulfilling it men were fulfilling that elementary zoological law which the bees fulfil by exterminating each other in the fall and according to which makes animals exterminate each other. No other answer can be given to this terrible question.

*Leo Tolstoy, 'Tales of My Army Life'*

2 | Sociobiology hopes to find laws that might explain certain kinds of human behaviour. Tolstoy wondered if humans kill each other because of some law of nature; perhaps the same law of nature that explains why bees kill each other in the autumn.* Sociobiologists today might ask what group violence and kinship among chimpanzees tell us about human behaviour. Reductive determinism can be immensely powerful, but such descriptions are at bottom limited by the degree to which we believe we can be reduced to, say, bees or chimpanzees.

Science provided answers to thousands of extremely clever and ingenious questions of criminal law, but it had no answer to the one he was trying to solve. He was asking something very straightforward: why, by what right, does one lot of people lock up, torture, exile, flog and put to death other people, when they are no different from the ones they torture, exile, flog and put to death? Instead of answers he got arguments about whether man had, or has not, free will. Can criminality be determined (or not) by skull measurements and the like? What part does heredity play in crime? Is there such a thing as congenital immorality? What is morality? What is insanity? What is degeneracy? What is temperament? ...

*Leo Tolstoy, Resurrection*

---

* In fact, it is rare for animals to kill their own en masse, and it is not generally true of bees, as Tolstoy thought; but it is true of a species of bee called Dawson's bee (*Amegilla dawson*) that lives in the hot soil of the Australian outback, and that could not have been known to Tolstoy. The male bees become such frenzied killers that by the end of the mating season every male is dead. Females that get in the way perish too.

[Tietjens] fell to wondering why it was that humanity that was next to always agreeable in its units was, as a mass, a phenomenon so hideous. You took a dozen men, each of them not by any means detestable and not uninteresting ... you formed them into a Government or a club, and at once, with oppression, inaccuracies, gossip, backbiting, lying, corruption and vileness, you had the combination of wolf, tiger, weasel and louse-covered ape that was human society.

*Ford Madox Ford (1873–1939), Parade's End*

It is a general rule of human nature that people despise those who treat them well, and look up to those who make no concessions.

*Thucydides (c.460–c.395 BC ), ancient Greek historian*

What does it all mean? Why did it happen? What made these people burn down houses and kill their own kind? What were the causes of these events? What force made people act that way? These are the involuntary, simple-hearted, and most legitimate questions that mankind poses for itself ...

*Leo Tolstoy, War and Peace*

3 | Psychologists, behaviourists, neurobiologists all want to know the answer to the same question: what is the basis of behaviour?

**B: If we could describe the behaviour of a fly, perhaps one day we will be able to describe the behaviour of all animals**

1 | The behaviour of the fly has been studied for over a hundred years.

2 | Richard Feynman once admonished the biologist Seymour Benzer (1921–2007) for telling Feynman's son that there are 100,000 transistors in the fly's brain. 'No, no. Tell it straight. They're not transistors, they're neurons. Don't oversimplify.'

3 | Behaviourism was at first purely a branch of psychology, but from the 1960s it began to be understood genetically. This radical shift was made by Seymour Benzer. Like Francis Crick, Benzer was inspired by Schrödinger's 1940s book *What is Life?*, in which the physicist wonders if life might be reducible to molecules. It could, and Crick was one of the discoverers of *the* molecule, the DNA molecule. At Caltech in the 1960s, Benzer began to look for reductive explanations of behaviour at the scale of the gene.

Behaviour is 'the way the genome interacts with the world'.
*Seymour Benzer*

4 | Benzer chose the fly. The fly's behaviour – at larger scales – had already been studied extensively for over half a century. He called flies 'atoms of behaviour'.

5 | Benzer was a founder of a new discipline called behavioural neuro-genetics, which, as the words tell us, is a mash-up of behaviourism, neurobiology and genetics. Behavioural neurogenetics looks for explanations of behaviour at the level of genes: particularly neurologically disordered behaviour associated with mutant genes. Behaviourism itself is largely observational. Behavioural neurogenetics offers an explanation at the molecular level for observations made at the macroscopic level. Today researchers are also looking for behavioural explanations at the cellular level and at the level of neural circuits.

6 | Benzer and others showed that a range of different behaviours – the ability to learn and remember, courtship and sexuality among them – are controlled at the genetic level.

7 | Benzer developed a technique called chemical mutagenesis for investigating the genetic make-up of his flies. Chemical mutagenesis is a process by which the genetic information of an organism is changed in a controlled manner using chemicals, typically ether. It is a very

precise technique. It is not uncommon to change a single nucleotide of the DNA molecule at a time. Populations of these mutagenised flies are bred and laboriously scrutinised for visible evidence of the artificially manufactured mutations, which might typically be aberrant behaviour, or some physical deformity like a missing wing.

8 | The first result from Benzer's laboratory was to show that the fly's biological clock* is controlled genetically. Using mutagenesis, a population of flies was screened, looking for behaviour in which a particular fly's biological clock had become corrupted. These flies were then isolated in order to identify the mutagenised gene thought to be responsible for the aberrant behaviour.

9 | In the early 1970s, Chip Quinn, one of Benzer's post-doc researchers, discovered that he could teach a fly. If exposed to an odour at the same time as it receives an electric shock, the fly learns to associate the two, and so will avoid the odour in future, even when the electric shock is withdrawn. This is called adaptive olfactory learning. Quinn chose two smells, one like licorice and another 'a lot like tennis shoes in July'. Repetitive training of flies can create long-term memories, which in flies means several days. This learning technique is still a powerful tool, used today in the hundreds of departments around the world investigating flies' brains. In 1976 Chip Quinn identified the first olfactory learning mutant gene – named *dunce*.

10 | The entomologist Vincent Dethier (1915–93) wondered for years if there was someone inside the fly who can learn. In those early days of behavioural research it was thought that all fly behaviour was inherited: that flies were automata. It is only in recent decades that invertebrates have been understood as simplified versions of humans, rather than merely models of themselves. By treating flies as if they are machines

---

* All fruit flies take an afternoon siesta. How they knew when to nap was a mystery.

we have discovered that they are not machines, or, if still a machine, more complex than the machine we first had in mind.

11 | The fly's behaviour can be uncannily like our own. When flies inhale alcohol fumes they first stagger and lose coordination, and then fall down. If chronically exposed to alcohol they become less sensitive to the effect. Male flies repeatedly rejected by female flies are more likely to turn to alcohol.

12 | Many genetic mechanisms that describe adaptive behaviour, as well as key components that describe how memory and learning occur, are conserved across the animal kingdom. For example, a protein (named CREB), involved in the process of translating short- into long-term memory in the sea snail *Aplysia*,* serves much the same function in the fly *Drosophila* and in mice. In humans the protein plays a significant role in drug addiction.

13 | At one time the idea that a single gene might control aspects of behaviour and learning was something no one wanted to believe. Now it is thought that even single neurons might impact on behaviour: reductive materialism at its most powerful.

**C: If the fly's brain is a machine, perhaps all brains are machines**

1 | These days scientists are able to study the fly at the level of the neural circuit, a scale up from descriptions at the genetic level.

2 | A fairly new and powerful reductive belief is that the nervous system is built out of many repetitions of a limited number of different types of neural circuit, and that within the great complexity of the brain are a limited number of fundamental (or so-called 'canonical') circuits

---

\* The biologist Eric Kandel spent years teaching *Aplysia*. Knocking it on the head makes the snail forget, just as it does to humans.

which perform fundamental operations that are preserved across all complex living creatures. It is assumed that during the course of evolution, complex organisms like humans and kangaroos have retained some of the mechanisms of learning and memory that can be more readily investigated in simpler animals, like worms and flies. It is not yet known for sure if there are such things in animals as canonical circuits (preserved, as genes are, across all living forms), what they look like or what they do. In ten years there may be answers to some of these questions. If we can find out how intelligence arises out of the physical interaction of nerve-cell components that are themselves unintelligent and can be modelled, then there is some hope that down the line either the same approach will be helpful in our understanding of consciousness, or that out of our deeper understanding of intelligence we will understand consciousness in some different and unforeseen way that will make it clearer what to do next.

Ironically, the strongest evidence that neurons are arranged in circuits comes from electronics and machine modelling of intelligence. Ironically because machine learning was inspired by biology in the first place, even though the biological systems themselves were poorly understood. In the world of electronics, much effort has been put into trying to model intelligence. For now, since scientists do not have a reductive model to describe intelligence in the brain, they have hijacked the electronic model.

3 | i In electronics, intelligence is broken down into a number of modules: a data storage system, an address system that activates the data storage locations for reading and writing information, a data input line, a data output line and so on.

ii If the brain is like a piece of electronics, there are many questions to answer, for example: How are neurons organised into circuits? How do these circuits process the sensory information that floods our bodies from the external world? How do these circuits store memories in the brain? How do these circuits retrieve memories?

4 | At its most reductive, behaviour is all about what do I do next? How do humans, or how does any intelligent organism, decide what to do next? How do we calculate what makes a good choice or a bad one? How do we learn?

5 | From a reductive Darwinian point of view, intelligence has evolved out of the need for a moving organism to know what to do next if it is to survive in a changing environment.

6 | Even bacteria show signs of intelligent behaviour. A bacterium like *E coli* possesses a propulsive tail, called a flagellum, which is activated by chemical sensors that are in turn controlled by a biochemical circuit. It is the signalling of this circuit that allows the bacterium to search out nutrients and to avoid toxins. *E coli* may show basic intelligence, but it does not learn. Sophisticated intelligence is the ability to predict the consequences of the actions we take. That ability relies on learning and memory. An animal learns to pay attention to a stimulus when it is important to its survival to do so.

7 | It takes only three visits of a bee to a flower for the bee to be taught to prefer a yellow flower over others. The bee has no long-term memory, yet complex thinking can result out of the connectivity of a modest number of neurons.

8 | At the machine level, learning is about the association of two stimuli, and memory recalls that association. Learning is a process that first measures the gap between expectation and experience and then subsequently narrows the gap. A learned experience is one in which experience is in agreement with expectation. Learning begins as some kind of measurement, an error signal. When the error signal has been reduced to zero the learning process has come to an end.

9 | It would seem that the old adage about learning from our mistakes is true. Trial and error is a part of learning, but not all. In order to learn

from our mistakes we must somehow be able to measure our improvement over time. Learning is a process that continually alters the action taken, even though the stimulus remains the same. And it is a process that has an end. Somehow we must remember that we have learned something. In this way learning and memory must be intricately tied together.

10 | It would be a huge breakthrough if we could find a learning circuit even in a fly, not just in our understanding of a fly but in our understanding of the neural systems of all animals with a nervous system. Understanding how neurons work together to generate thoughts, memory and behaviour is one of the most difficult problems in biology. It is assumed that a learning circuit is some kind of closed loop: a bootstrapping system comparable to a feedback circuit in electronics. In electronics, error signals are integral to many feedback systems.

11 | There are no measuring instruments yet developed that can record the activity of a complex neural net, particularly as part of a living human brain. For now, researchers look for simple circuits in creatures with simpler brains. The brain of a fly is a network of some 200,000 neurons, compared to 300 million neurons in the brain of an octopus, and 100 billion in a human.

12 | Most learning experiments that are carried out in *Drosophila* today are still based on adaptive olfactory conditioning – an association of an odour and a punishment* (typically an electric shock). Punishing a fly causes dopamine to be released by neurons in the fly's brain. Dopamine is a hormone and a neurotransmitter – a chemical messenger that transports information from one cell to another. Tiny amounts of the chemical alter the cell's metabolism. Adaptive olfactory learning in a fly is analogous to dopamine learning in humans, except that in humans dopamine-release signals a reward rather than a punishment.

---

* A reward system is also effective, but implicates a different type of neuron.

Reward and punishment systems maximise survival in animals, whether finding food, responding to predators, or identifying a mate. The fly olfactory system reproduces in miniature all the essential features of our own, including an ability to discriminate and learn. Current research is trying to work out the role dopamine plays at the neural level in the fly when the fly learns some new behaviour.

13 | Out of the 200,000 neurons in a fly's brain only about two hundred are dopamine-producing neurons. By splicing in part of the genetic code taken from naturally phosphorescent creatures, like certain kinds of jellyfish,* it is possible to label the dopamine neurons; that is, they light up when the neurons are active. This technique has made it possible to narrow down which neurons are associated with a particular behaviour simply by looking for the lit-up neurons under a microscope.

Conversely, a technique has been developed that allows specific neurons to be activated at a distance using light (in the form of a laser). This new discipline is called optogenetics, and was invented separately by Gero Miesenböck and Karl Deisseroth. Simply by pointing a laser at the head of the subject, typically (for now) a fly, a light-sensitive gene is activated. The gene makes a protein that causes an ion channel to open and the neuron to fire.

14 | In the early years of this century Susanna Lima (Gero Miesenböck's then research student) identified the dopamine-producing neurons – there are just two of them – which fire when a fly tries to escape from, say, a predator. Lima marked these neurons genetically so that she could activate them artificially by shining a light on them. The fly flew

---

* Soon after the discovery that the gene from light-emitting organisms could be added to the genome of those without this ability, some scientists began to experiment seemingly for the fun of it: a tobacco plant that glowed in the dark, bacteria that flashed on and off like Christmas-tree lights, and most spectacularly, a fluorescent albino rabbit, named Alba.

away even though there was no predator nearby. To ensure that the fly was not responding to the looming presence of the researcher, she performed a brutally simple experiment. She cut off the fly's head. Surprisingly, a headless fly can survive, inertly, for a day or two. A fly's escape mechanism is housed not in its head but in its thorax (the rough equivalent of a vertebrate's spinal cord), so even if the fly's head is cut off, its escape mechanism is retained. When the light was shone on the headless fly, it dramatically (and, without a head, clumsily) flew off.

> When we saw the headless bodies flying away, we were absolutely stunned.
> *Gero Miesenböck, biologist*

15 | In 2007 Karl Deisseroth and his Stanford colleagues used an optical fibre to deliver light directly into the brain of a mouse to activate neurons identified in the control of sleep. They were able to wake sleeping mice.

16 | In Canada in 2009 the memory of a frightening noise was erased from the brain of a mouse. The memory was traced to specific nerve cells that encoded it.

17 | It is known that hungry flies are better at learning than flies that are replete. In 2009 Scott Waddell and his colleagues identified six dopamine-producing neurons which when stimulated reduce hungry flies' motivation to learn. They appear to have isolated a 'master switch' that controls some aspect of a fly's motivation. It is motivation that gives behaviour purpose and intensity. Motivation has interested psychologists and animal behaviourists for decades, and researchers are now finding explanations at the level of circuits of neurons.

18 | Male courtship displays are elaborate in many species. Male fruit flies vibrate one wing in a way that female fruit flies find irresistible. By activating the neuronal circuitry responsible for this behaviour, females

can be induced to perform the same behaviour. In the wild such behaviour is almost exclusively male. It would seem that male and female brains are wired in the same way, but that differences in sexual behaviour arise out of the action of specific master switches that set circuits to either male or female mode. Why do females retain neural circuitry dedicated to a behaviour they never express? One possible explanation is that the circuits involved in courtship are held in common with other programs, like, say, motor programs that control the movement of the fly's wings. A largely unisex system containing high-level switches saves the organism from having to build separate nervous systems from scratch according to some different male or female plan. Finding master switches and basic circuits could revolutionise biology in the way that the discovery of DNA did.*

19 | The vast majority of biochemical reactions that govern the behaviour of cells and organisms are controlled by types of proteins called regulatory proteins. If these proteins could be rewritten so that they were made light-sensitive, it would be possible to control the workings of the cell remotely.

20 | Sociability, common sense, altruism, empathy, frustration, motivation, hatred, jealousy, peer pressure ... It is conceivable that we might one day describe the emotions of a fly.

> In the past, questions like what is memory, motivation, attention and decision-making have been answered only at the psychological level. It is revolutionary to be asking these questions and looking for answers at the neural level.
> *Gero Miesenböck*

21 | Vision begins as photons of light striking the retina, hearing is a response to sound waves, and smell is the recognition of certain

---

* See page 161, 7.

molecules, but perception is more than simple acknowledgement of these signals and substances: our senses are themselves mediated by learning and memory. It may be too early to make any progress in the understanding of perception at a reductive level in the brain, but we know that perception is somehow mediated by learning and memory, so if we can begin to understand memory and learning that is progress.

We look at what is in front of us. It's not that we're oblivious to the rest, but for neuroscientists consciousness comes way down the list of concerns. I'm looking for descriptions that take us from sensation to action. All the rest is philosophy and poetry.
*From a conversation between a neuroscientist and the author*

22 | Behaviour, like the brain, the universe, consciousness and human beings, is a problem that requires descriptions at many levels, black box within black box.

# SECTION 9

# There is always something missing

I can calculate the motion of heavenly bodies, but not the madness
of people.
*Isaac Newton (1642–1727)*

We are fearfully and wonderfully made.
*Psalm 139:14*

Who can calculate the orbit of his own soul?
*Oscar Wilde (1854–1900), playwright*

A continual allegory – and very few eyes can see the Mystery of his
life.
*John Keats (1795–1821), poet*

I have never been able to believe that any system, no matter how
seductive, can hold the ambiguities that are inherent in being a
person in the world.
*Siri Hustvedt, The Shaking Woman*

Information underrepresents reality.
*Jaron Lanier*

1 | In 1957, Julian Huxley coined the word transhumanism, the idea
that we can use technology to transcend our bodies. For those of us

who believe human beings are already transcended flesh, it would be hard to say what can be transcended, unless we are to become some kind of transcendence of transcendence.

> I think in years to come we'll be able to download our personalities onto computers and have them live on in virtual worlds after we die. Then our consciousness will survive death.
> *Kevin O'Regan*

2 | The futurist Ray Kurzweil believes that before 2050 the brain will be copied and uploaded into a non-biological device.

3 | Gordon Bell, a Microsoft computer scientist, now in his seventies, records everything he can of his life. He wears a SenseCam camera around his neck in order to capture what his eyes are seeing. Every phone call is recorded, every piece of paper he reads is scanned into his computer.

4 | The physical world is ultimately reducible to yes or no answers. In 2001 Seth Lloyd, professor of mechanical engineering at MIT, calculated the number of possible computations that have taken place in the universe since the Big Bang as $10^{120}$. In 2004 cosmologist Lawrence Krauss calculated that the universe had the same number of computations left to perform.* But are human beings reducible to computers? The brain is dripping with information, but it seems that a self cannot be extracted from it that is made of bits of information. The self appears to be some kind of interaction of the body and the environment that gives information meaning.

5 | The Hubble telescope stared at a single spot in the sky for twenty-two days; except that it does not stare, it merely records information. Humans stare as they pore over the information that the telescope has

---

* ($10^{40 \times 3}$). Discoveries that would have pleased Dirac and Eddington. See footnote p.46.

collected, and make out of it what we call sense. The mind is not a camera. Seeing is not enough. There has to be thinking and meaning.

Seeing is not just about looking, but directing the mind's attention to what is being looked at. Buddhists call it mindfulness, the act of being present and attentive. Our scientific instruments reveal what we might not otherwise see, but first a human being must see what has been recorded, and then interpret it.

6 | Paul Cézanne sometimes stared at a single apple for hours.

> When Picasso had looked at a drawing or a print, I was surprised that anything was left on the paper, so absorbing was his gaze. He spoke little and seemed neither remote nor intimate – just completely there.
> *Leo Stein (1872–1947), art collector and critic, brother of Gertrude Stein*

7 | One night in the 1950s, a night when the University of Chicago's telescope was open to the public, one of the visitors told the astronomer on duty, Elliot Moore, that she could see a star in the Crab Nebula that was flashing. Moore told her that the star could not be flashing, and explained to her the difference between twinkling and flashing. The woman replied that she was a pilot, and knew about scintillation. She insisted that the star was flashing, not twinkling. Moore recalled the visitor (whose identity remains a mystery) years later. In 1968 it was discovered that there are pulsing stars near or coincident with the Crab Nebula, and that one of them – now called the Crab Pulsar – is at the heart of the nebula. The Crab Pulsar is a young neutron star left over from the explosion that we witness as the Crab Nebula, and which was first seen on earth in 1054. The Crab Pulsar pulses thirty times a second.

8 | The astrophysicist Jocelyn Bell spotted a regularity that no one else had spotted in an otherwise seemingly random radio signal from outer space.* It proved to be the first evidence of quasars.†

9 | Whatever kind of machine we are as humans, we remain, for the moment at least, more complex than our best technologies. In our determination to prove that we are machines, we are ourselves in danger of becoming *mere* machines. When humans are animals we are mere animals, when we are robots we become things, and possibly mere things. As Alan Bennett once said, when we say that men die like flies that is exactly what they become, like flies. To save our humanity we can elevate ourselves or we can elevate the fly, or both. If we do not have it in us to elevate the human, we might elevate the machine, and with much the same effect. It seems likely that we will always be able to make better tools and find better analogies for what we are. If we believe that we can make better tools forever, it is not so big a step to believe that our understanding of what flesh is may be forever beyond our reach too.

> A new generation has come of age with a reduced expectation of what a person can be, and who each person might become.
> *Jaron Lanier*

---

* For a moment it looked like evidence of aliens, but when it comes to it scientists don't really believe in aliens, and even the possibility of their existence is a spur to search for some other plausible explanation. The possibility of human specialness and seeming evidence of the existence of God are similar spurs. It's a good working strategy, but it doesn't mean that any or all of these possibilities are ultimately excluded, just doubted for as long as doubt is possible.

† On the other hand, the astronomer Percival Lowell (1855–1916) saw criss-crossing lines through his telescope that he believed were evidence of canals on mars. Later he could see the same lines on photographs. The photographs were even published, but no one else could see the lines he could see.

10 | The philosopher and neuroscientist Sam Harris says humans might eventually have to sacrifice themselves to a higher intelligence; presumably because intelligence somehow trumps being human. Why do we privilege intelligence among the qualities that make us human? What if we were able to make robots more loving, more moral, or more conscious than we are?* What if we knew how to make robots more human than humans; if we but knew, of course, what being a human being entails. Losing to the chess computer Deep Blue in 1997 was maybe greater evidence of Gary Kasparov's humanness than of Deep Blue's intelligence. It is not clear, in any case, that we have got very far in our attempts to manufacture intelligence. AI (artificial intelligence) is more about finding ways to mimic intelligence than it is about creating it from scratch.

11 | We do not feel less ourselves for knowing we are merely atoms, so why should we feel lesser for being reduced to machines or to the scheming selfish entities of sociobiology and evolutionary psychology? There is always something missing in any reductive description, no matter how powerful it might be.

12 | Reductionism always looks like a fundamental description, because without the atom, gene, cell, neural circuit, body, population, there is nothing. Remove the midbrain and eye movement is not possible. Remove the oblongata and breathing is not possible. Remove the atoms and there is nothing.

There are some processes that seem to be irreducible. Blood clotting, for example. Remove various proteins involved in blood clotting and blood will still clot. Remove others and it will not. The process appears to be irreducibly complex and mysterious, at least for now; such examples are held up as hopeful evidence by transcendentalists. If this is the best they have, it is not enough.

---

* The days when we lived close to other *Homo* species are long gone. Might we now begin to look forward to days living alongside our successors?

13 | Tracing the universe back to the Big Bang is rather like tracing a river back to its source. There may be some highest patch of damp ground from which the river may be said to originate, but this is to ignore everything that pours into the river along the way. In the story of the evolving universe, what pours in is scientifically unpredictable transcendency, what scientists call emergence.* Descriptions of a single molecule of water do not predict its possible states as ice, liquid and steam, let alone its myriad other properties. As the universe unfolds, so we need different kinds of description at different scales.† The predictive powers of science working up from the bottom are restricted. We would predict very few features of this universe – not the arrival of human beings, certainly – if all we had to go on was particle physics.

The physical world can be approximately reduced to what appear to be laws of nature. But just because we can build the edifice of the universe on these laws, that does not make the laws more real than what emerges later. If human beings are a kind of illusion from the perspective of physics, the illusion *is* our reality. God, Nature, being human, are all forms of transcendency, or emergence, which is, I believe, not so much a spiritual statement as a logical one.

What is surely impossible is that a theoretical physicist, given unlimited computing power, should deduce from the laws of physics that a certain complex structure is aware of its own existence. *Brian Pippard, physicist*

Moving from the sensory world to the internal world gets harder and harder. In physics 'proof' is in the measurement, i.e. the number of decimal places. Proof in biology doesn't look like that. *From a conversation between a neuroscientist and the author*

---

* What David Deutsch defines as 'explicability at a higher level'.

† See *Knocking on Heaven's Door* (2011) by theoretical physicist Lisa Randall for a detailed exploration of the significance of scale in scientific theories.

14 | What would an alien scientist make of a piano? It is a mechanical object, a percussion instrument. Sounds are made out of hammers striking strings. The sound can be modulated by how hard and how fast the keys are depressed. Given that the action of the piano is fixed, it might even be true that the only variable is the force with which the keys are struck. The piano is little more than a glockenspiel with pedals. Would it be possible to predict what a history of performance there is, or what the piano has evolved into? The interaction between machine and performer is the triumph of art and artistry over artifice. No student of the piano would get very far if she only considered the physical constraints of the instrument. It has to be approached with the human qualities of love, tenderness, fierceness, and so on, not merely controlled force and velocity.

The behaviour of high-level physical quantities consists of nothing but the behaviour of their low-level constituents with most of the details ignored.
*David Deutsch*

# When the gene is no longer enough

1 | Every reductive scientific description is circumscribed. There are limits beyond which the description either breaks down or is ineffectual. Life, for example, does not care about sub-atomic particles or quantum mechanics. The bottom limit of life appears to be the DNA molecule. At its most reductive, life is the story of what happens to DNA as the so-called 'target' of natural selection.

2 | Natural selection as the mechanism underlying adaptive change (with the gene as its target) is an idea brought to popular attention and powerfully defended by Richard Dawkins, first in *The Selfish Gene* (1976).

3 | For many biologists the gene is not the only target of natural selection. Stephen Jay Gould writes that it is not the gene but 'a nested hierarchy of biological individuals (genes, cell lineages, organisms, demes, species, clades)'.* Gould believed that genetic determinism is just one kind of biological explanation. He believed that there is a hierarchy of explanations at different scales, not just at the scale of genes but at the scale of individual organisms, species, population explosions and mass extinctions, even of the earth itself. Ernst Mayr† also denies the gene as the unit of evolution. Mayr says that it is neither the gene as sole target,

---

* *The Structure of Evolutionary Theory*, published in 2002, the year he died.

† In *What Evolution is* (2001).

or the individual as Darwin thought, but different targets at different scales.

> I reject the reductionism that underlies [the] search for general laws of highest abstraction.
> *Stephen Jay Gould (1941–2002), biologist*

4 | Niles Eldredge* argues that Dawkins's view works best as an explanation of generation-by-generation change within populations, but is 'useless as a general evolutionary theory covering the large-scale events in the history of life', just as large-scale changes – like a change in climate – are hopeless at describing the 'internal dynamics of selection on smaller scales'.

It is hard to explain, taking the gene as the target of natural selection, why so many creatures remained unchanged for millions of years. The nautilus, for example, was around for fifty million years before its brain size suddenly increased in competition with bony fish.

5 | The philosopher Mary Midgley has described the use of the word 'selfish' in the theory of evolution as 'the most glaring example of thoughtless use of metaphor over the past thirty years'.

6 | Controversially, cognitive scientists Jerry Fodor and Massimo Piattelli-Palmarini have argued that much of evolution happens because of physiological and developmental constraints, not out of natural selection. The reason pigs do not have wings is not because winged pigs lost out to wingless ones: there could never be pigs with wings. What the nature of these possible constraints is has yet to be

---

\* In 1972, with Stephen Jay Gould, he proposed the punctuated theory of evolution: 'that nothing substantial happens in terms of accruing adaptive evolutionary change unless and until physical events upset the ecological applecart, leading to patterns of extinction and evolution of species'.

elucidated. It may be that creatures with a particular trait also have other traits that are linked to it developmentally.

> The squid somehow knows, the wood duck knows, the unstated fractal rules out in nature know; there are consistencies, parallels, in the way patterns form, and ever more human attention is needed to grasp them.
>
> *David Rothenberg, professor of philosophy and music*

7 | A certain type of gene, called the body-plan gene or Hox gene, at first sight appears to be an argument against natural selection. These genes when removed from one creature and inserted into a different kind of creature behave as they should for the appropriate animal. Insert, for example, the mouse version of the Hox gene pax6 into a fly, and the fly makes fly eyes not mouse eyes. If these genes were selected by natural selection it must have happened early on in the evolution of life. But then, evolution was happening at the molecular level for billions of years before the explosion onto the scene of organisms larger than bacteria. There is still plenty of room for explanation at the bottom.

8 | Much of evolution may be constrained for developmental reasons. Many genetic changes may come along for the ride as a result of some selected genetic change, but how to tell which are the naturally selected changes and which the hangers-on? It may be impossible to say. There are features that seem to come as part of a package. It is that entire package that evolves by natural selection. Male nipples are an example of such a hitchhiker. Nipples offer no adaptive advantage to men, they merely come as part of a common developmental pathway that males share with females.

Evolutionary-genomic studies show that natural selection is only one of the forces that shape genomic evolution and is not quantitatively dominant, whereas non-adaptive processes are much more prominent than previously suspected.
*Eugene Koonin, evolutionary biologist*

There is often a lack of hard evidence to support much of what is claimed to be the result of natural selection.
*John Endler, evolutionary biologist*

9 | Stephen Wolfram* and others are in search of a new 'atomic' unit of biology to explain some of the puzzling features of the natural world. Wolfram has argued that much of the relatedness and patterning of nature arises out of the iteration of a few basic rules applied repeatedly to some simple cellular automata. For him, much of nature is the result of the turning of some fundamental machinery rather than the outcome of the randomness of evolution.

While natural selection is often touted as a force of almost arbitrary power, I have increasingly come to believe that its power is remarkably limited.
*Stephen Wolfram, scientist*

Natural selection is only a filter and filters cannot be the sole cause of the coffee that comes from them.
*Mary Midgley, philosopher*

---

* Maverick genius, designer of the computational program Mathematica, author of *A New Kind of Science* (2002).

Evolution doesn't really explain much.* It's more a way of describing the context of history. There is an unlimited potential for generating little machines in nature, they are built on each other's shoulders, but they are not really evolving in a way that makes much sense.

*Ofer Tchernichowski, neuroscientist*

10 | Several evolutionary advantages to walking upright have been put forward. It lifted the head away from the hot ground, it meant we could survey for predators more easily, it allowed us to throw spears. The problem is that we cannot know if this trait was selected genetically as an adaptation, nor for what reason, nor point at which genes were involved.

11 | There are evolutionary biologists who will interpret almost every facet of biological life in strictly adaptationist terms. The mechanism begins to weaken when the explanations become mere storytelling. Stephen Jay Gould called such explanations Panglossian: that all biological phenomena are adaptations for the best in the best of all possible worlds.

Many people today are infatuated with the biological determinants of things. They find compelling the idea that moods, tastes, preferences and behaviors can be explained by genes, or by natural selection, or by brain amines (even though these explanations are almost always circular: if we do x, it must be because we have been selected to do x).

*Louis Menand, New Yorker magazine*

12 | It has been argued that, from an evolutionary point of view, we have the psychological make-up that we have because our brains

---

* Cf: In 1973 the geneticist Theodosius Dobzhansky wrote, 'Nothing in biology makes sense except in the light of evolution.' Some biologists no longer agree.

evolved for evolutionary purposes that are now redundant but are nevertheless still wired into how we think. Once we needed to understand the weather in order to survive, and we learned to look for its patterns and causes; and so, the theory goes, we look and find those patterns still.

Evolutionary explanations that draw on our behaviour when we were hunter-gatherers are these days offered for almost every kind of contemporary human behaviour: from rape to body-fat content, from marital infidelity to altruistic behaviour.

13 | Neuroscientists Anya C. Hurlbert and Yazhu Ling constructed a test that shows that the colour pink is preferred by girls and blue by boys. The preference, it is argued, has arisen because when we were hunter-gatherers women were the gatherers of ripe fruit, and hunting men would have looked to the sky for good hunting weather. Critics have pointed out that in Victorian England the colours were reversed, blue for a girl and pink for a boy; which does not so much undermine the findings as question the explanation.

14 | In a study of 90,000 people in the UK aged between thirteen and ninety a marked preference was shown for Impressionism over Cubism. But what does this tell us? It has been suggested that it is because Impressionist paintings are more in tune with the amygdala. The amygdala is our early-warning system, on the lookout for blurry forms moving across our peripheral vision. Another study, however, claimed to find a preference for abstract art, and argued that this is an adaptation from our deep past when we swung through trees. There was a survival edge to those best able to make a connection between the angle of branch and the angle of arm.

15 | I read that people with more symmetrical faces are more self-sufficient. Other studies have shown that people with more symmetrical faces suffer fewer congenital diseases, and so make better mating partners. In tests, people with more symmetrical faces were less likely

to cooperate, and less likely to expect others to cooperate. It is argued that because people with symmetrical faces tend to be healthier, and seen as more attractive, they have less incentive to help others: 'Through natural selection over thousands of years, these characteristics continue to the present day.' These kinds of studies, the researchers suggest, might 'help to design public policies and act as a corrective to purely economic-based decision making.'

16 | There are many theories meant to account for the evolution of language. Language evolved between 100,000 and 150,000 years ago, perhaps more recently. The human larynx had dropped by this time, which made it easier to choke but also easier to speak and to make more sounds,* so if the evolutionary pressure was adaptive, the advantage must have outweighed the disadvantage of being more prone to choking. Another theory tells us that men acquired social status for their public speaking, and with social status comes greater opportunity for sex and reproduction.

17 | In a paper entitled 'Why do Gentlemen Prefer Blondes?'† the neuroscientist V.S. Ramachandran argued that evolution explains why. Paler women more easily show signs of disease, which dark skin can disguise, so by choosing pale women men can be more certain that they

---

* The larynx drops in humans between the ages of three and four; before then, infants can breathe through the nose and swallow milk at the same time. Chimpanzees have the wrong architecture of the mouth and throat for speech. They can sign, but only if taught by humans. Language is innate in humans, but not in chimpanzees.

† The sequel to Anita Loos' novel *Gentlemen Prefer Blondes* was titled *But Gentlemen Marry Brunettes*. In the introduction, Loos wrote: 'Recently ... the question was put to me "... If you were to write such a book today, what would be your theme?" And without hesitation, I was forced to answer, Gentlemen Prefer Gentlemen (a statement which brought the session abruptly to a close). But if that fact is true, as it very well seems to be, it, too, is based soundly on economics, the criminally senseless population explosion which a beneficent nature is trying to curb by more pleasant means than war.'

mate with a healthy specimen. The paper was in fact a hoax meant to show the limits of adaptationist explanations.

18 | Richard Lewontin has pointed out that as no one has yet managed to associate a particular set of genes with any human behaviour, 'all statements about the genetic basis of human social traits are purely speculative'. Yet biological explanations abound. It has been argued as an explanation of religion, for example, that because indoctrination is easy to accomplish in humans, religion confers a biological advantage by giving apparent meaning to life.

19 | Arguments that retrospectively explain some physical structure by reference to its supposed fitness in some hypothesised environment of the past are used in fields as diverse as philosophy, psychology, anthropology, sociology, and even aesthetics and theology. Jerry Fodor and Massimo Piattelli-Palmarini call this type of explanation imperialistic selectionism. Mary Midgley calls it scientism: treating science as some kind of universal belief system.

20 | In his book *On the Origin of Stories* (2009), Brian Boyd writes that 'Art is a specifically human adaptation … It offers tangible advantage for human survival and reproduction, and it derives from play, itself an adaptation widespread among animals with flexible behaviours.' But as Raymond Tallis points out, it is often hard to see what these advantages might be. 'An organism that devotes many hours to the solitary pastime of reading, and reading-inspired daydreaming, would surely be less fitted for the hurly-burly of everyday life than one satisfied by the one-sentence paragraphs of the tabloid newspapers.'*

21 | Barbara and Daphne were two of the participants in a study of identical twins that took place in the 1990s. The twins had met at King's Cross station after being separated for four decades. They were both

---

* Raymond Tallis, *Aping Mankind* (2011).

wearing beige dresses and brown velvet jackets. They both raised a crooked finger in greeting. They discovered they both liked their coffee black, and that blue was their favourite colour. They were both sixteen when they married, and both laughed more than anyone they knew. That the twins have identical genes can hardly begin to explain how they could be wearing the same clothes on the same day after being apart for four decades. An environmental explanation can fare no better.

22 | Darwin could not work out why there was so much beauty in the world. He once wrote that 'the sight of a feather in a peacock's tail, whenever I gaze at it, it makes me sick'. He introduced sexual selection as another mechanism, alongside natural selection, to explain natural adornments, like the peacock's tail, that clearly offer no survival advantage. Rather the opposite, a male peacock's enormous and brilliantly coloured tail impedes its ability to survive.

Sexual selection shows that when females are in charge, beauty ensues. Females decide for apparently aesthetic reasons what genes get to be duplicated, not because of fitness but out of apparent whim.

To account for the male peacock's tail, the biologist Amotz Zahavi has argued that it proclaims: 'Look how strong I am. I have all this power to spare because I can still bear this huge useless tail.' It has been argued that the nests of bowerbirds, with their decorations of seashells and berries, are meant to signal to the female the bounty and the healthy diet of the male. Others have described the nests as possible anti-rape structures that offer females the opportunity to consent. But this is not how Darwin saw it. For him the choice was determined by what 'delights the mind of the female'. Darwin realised that what females choose in a mate is often arbitrary, and has no distinct feature except that the females presumably like their mates that way. If Darwin is correct, then what delights the mind of, say, the peahen or the female bowerbird presents a measurement problem. How do we decide which is the most beautiful peacock tail, or the most beautiful bowerbird's nest? Sexual selection brings an element of *sprezzatura* into biology. It is not the most useful or even the most interesting trait that survives, but what most amuses. It also turns evolution inwards, away from the external world and into the mind of, say, a bird. Biology becomes aesthetics. But for adherents of the gene such as Richard Dawkins, 'nature cannot afford frivolous *jeux d'esprit*. Ruthless utilitarianism trumps, even if it doesn't always seem that way.'

23 | According to the ecologist Richard Prum, the huge diversity of life is simply everything that is possible. He says that in order to avoid facing the 'mysterious arbitrariness of nature', we are in danger of ignoring 'almost everything interesting'.

24 | So that they do not stray into the realm of aesthetics, adaptationists are not only forced to ignore the possibility of beauty for its own sake, but in order to preserve an adaptationist explanation they construct stories that stretch credibility.

> All sorts of stories can be told. Not all of them need to be believed.
> *Mikhail Bulgakov (1891–1940), The Master and Margarita*

25 | Peter Osin, a consultant breast pathologist, shows how fraught the process of interpreting even scientifically-collected data can be:* 'One study … offered the conclusion that women who have had cosmetic breast augmentation are between two and three times more likely to commit suicide than the ordinary population of women … But another study used similar data to suggest that breast augmentation actually *lessens* the risk of suicide. So which is right?' The first study compared the results to the general population of women, the other only to those women who would consider having cosmetic breast augmentation. Women within this subset of the female population have been shown,

---

* Writing in the Royal Opera House programme to the opera *Anna Nicole* (2011). Composed by Mark-Anthony Turnage, the opera is based on the life of Anna Nicole Smith, a model famous in part for her augmented breasts.

Osin writes, 'to display more alcohol and tobacco use, they have more numerous sexual partners and higher divorce rates and they place greater reliance on oral contraceptives and psychotherapy. Based on these criteria, one could expect to see a five-fold increase in the suicide rate among this group.' But the suicide rate among those who actually go ahead with the augmentation process is *only* two to three times that of the general population; so among women in this smaller population the suicide rate goes down. How you tell the story matters.

26 | When J.B.S. Haldane was asked what might disprove the theory of natural selection, he said: 'Fossilized rabbits in the Precambrian.' Biological proof exists within the context of plausible storytelling. Biological descriptions break down both when biology is pushed too far as a physical science, and when the stories become too far-fetched.

27 | You could run the universe over and over out of some physical description and always miss the bit where life came and went (if it comes and goes). Biological descriptions require a different focus. They are dependent on the storytelling techniques that humans have evolved. When biology found the explanatory power of the gene it became more like physics, and the need to tell stories was diminished. But when the gene is preserved at all costs, the stories can become too fanciful. The closer we get to the scale of human beings, the better the stories must become. Human beings tell the best stories about themselves, and they are not always stories about genes.

# SECTION 11

# On tools and human evolution

1 | The story of the survival of human beings is particularly difficult to tell as a story of adaptation in nature. How did the weakest ape come out so far on top?

> Man is of all others the most curious vehicle ... so slight a frame
> and so totteringly put together ...
> *Laurence Sterne, Tristram Shandy*

2 | Our skulls are so large we risk death even by being born. Our jaws are weaker than they once were, in order, perhaps, to allow the head to expand to accommodate a larger brain. Large brains need massive amounts of energy, which is why we need to cook food, because it makes energy available more easily. We are indeed totteringly made. The spinal column is better for suspending a ribcage than for acting as a column to support a heavy head. The change to a standing posture* resulted in the shortening of our intestines, which is why our digestion is relatively poor. We find it hard to keep warm, we overheat easily, we have fragile nails and teeth, a poor sense of smell, not to mention all the

---

* In the popular imagination the most famous image of human evolution is of a creature on all fours becoming increasingly upright and less hairy. This won't do. It is much harder to be bent half over than it is either to be fully standing or on all fours. However we came to be standing creatures it cannot have been through these intermediary stages. Perhaps it was the act of an individual – then imitated – an effort of conscious will and physical possibility.

senses that other animals have and we don't have at all. But the advantages of standing up are huge. Our hands are free to use tools, and to become tools. It was Aristotle who first pointed out that a hand is an instrument that represents many instruments. Standing upright also makes possible the kind of breath control that is necessary for speech.

3 | Non-human apes have a pelt, claws, and strong canines. An orangutan can kill a crocodile with its bare hands. What humans have instead is intelligence. Our brains are three to four times larger than any other ape's, far larger than they need to be to outsmart an ape.

4 | For three million years after we began to walk upright our heads expanded.* From an adaptationist's point of view it would be hard not to predict that once the head was teetering on a spinal column, the size of the head would decrease rather than increase. It would be a poor physical theory that couldn't make a predictive distinction between opposites, but biology isn't only a physical theory; it relies on a degree of storytelling based on the accretion of circumstantial evidence.

5 | The anthropologist Timothy Taylor (b.1960) has argued that the target of evolution moved from the gene to tools and technology. We are strong because of our wits and our weapons, and because we have made a life for ourselves indoors, protected from nature. Few of us now would live long in the wild, but even a puny human with a bow and arrow is king of the jungle. Skill became more valuable than brute force, for a while at least. To handle a longbow requires skill: any idiot can fire a gun, and many do. Technology allows us to outsource our intelligence and to share it. We don't need to know how a fridge works in order to benefit from it, and most of us don't. Today most of us are ignorant about the workings of most of our tools.

---

* These days our brains are shrinking. Compared to 10,000 years ago, our brains are 9.5 per cent smaller. We are also physically weaker, our eyesight is poorer, and our bodies are 7 per cent smaller than they were 10,000 years ago.

The popular lore of all nations testified that duplicity and cunning, together with bodily strength, were looked upon, even more than courage, as heroic virtues by primitive mankind. To overcome your adversary was the great affair of life. Courage was taken for granted. But the use of intelligence awakened wonder and respect.

*Joseph Conrad, Nostromo*

I wanted to beat the other boys and this seemed to be the way in which I could do so most decisively.

*G.H. Hardy, on becoming a mathematician*

6 | Human life seems to have broken the constraints of nature. At one time the fittest and strongest men won battles, but increasingly war became about strategy and weapons. The 'fittest' in an evolutionary sense became the smartest, not necessarily the strongest. An adaptationist explanation is preserved here if the unit of selection moves (far away from genes) to humans together with their weapons. Together, they take themselves outside nature, and in doing so evolution speeds up rather than slows down.

Our tools* enabled us to transcend selection by nature. We have been selected by our tools. We are their playthings. Tools allowed us to give up power for vulnerability. Tools allowed us to become gracile. We have used technology to create the environment in which we have evolved, and in which we continue to evolve at increasing pace.

Artefacts have souls and historical memories.

*Ertuğrul Günay, Turkish Minister of Culture 2007–13*

---

\* There is evidence of flint-napping about 40,000 years ago. Even that 'primitive' process involves a number of distinct steps in its evolution.

An insect-catching bird may have natural visual acuity hundreds
of times finer than ours, but we can track it, catch it, tag it, trap it,
or kill it at will … We can study its eyes to design new things.
*Timothy Taylor, The Artificial Ape*

7 | Humans never were part of nature. We were always part of techno-
logy. Humankind's first inventions were containers: bowls, jars, bags
and baskets.

Timothy Taylor suggests that one of the first containers that made
humans possible was the baby sling: 'we used technology to turn
ourselves into kangaroos'. The sling protected the fragile baby outside
the womb, allowing early, vulnerable, big-headed birth.

8 | Technology evolves a life indoors. We become increasingly cut off
from nature. Technology achieves what the playwright Max Frisch
called 'the knack of so arranging the world that it no longer needs to be
experienced'.

# SECTION 12

# Being rather than making

1 | It is not clear that we will ever be able to make a human being, nor ever understand its behaviour. But we might get closer to those possibilities if we could say what it is like to be one.

# PART THREE

# On Being Human

'What is a person?' If I knew the answer to that, I might be able to program an artificial person in a computer. But I can't. Being a person is not a pat formula, but a quest, a mystery, a leap of faith.
*Jaron Lanier*

# SECTION 1

# On culture

Humans woke from being organisms to being something quite different.
*Raymond Tallis*

1 | Once they were standing upright, humans became free to use their hands to point. A pointing finger* brings humans together in joint visual attention. Gesture, art and language indicate what it is to be another person.

2 | By putting their heads together humans 'have transcended biology'.†

3 | Between 35,000 and 40,000 years ago there was a creative outburst among humans. Tool use, military strategy, gamesmanship, morality, personality, religion, representational art, cooking, conversation, story-telling, jewellery, tattoos, ceremonial burials, jokes and laughter – some or all of these.

---

* Finger: from the Latin *fingere*, to shape, fashion or mould. Also the root of the word fiction, to make something up. On the ceiling of the Sistine Chapel Michelangelo's pointing finger of God is a pun.

† Raymond Tallis in *Aping Mankind* (2011).

Brains together create a space that cannot be stuffed back into the brain.

*Raymond Tallis*

We exist only to the extent that there are others.

*Johann Fichte (1762–1814), philosopher*

But can anyone doubt today, that all the millions of individuals and all the innumerable types and characters constitute an entity, a unit? Though free to think and act, we are held together, like the stars in the firmament, with ties inseparable. These ties we cannot see, but we can feel them.

*Nikola Tesla*

Every man has within himself the entire human condition.

*Michel de Montaigne (1533–92), essayist*

Our human world of pooled transcendence creates a theatre for our actions.

*Raymond Tallis*

4 | We live in the collective history made out of all of our individual and joint actions. We add to what has gone before, in art, conversation, jokes, cooking. Every human being is in debt in unknowable ways to everyone who has come before. Without shared context we would feel utterly alone. Not to feel alone is why we do the things that humans do: talk, laugh, dance, make things. Culture – whether it is art or conversation – maps our internal landscape onto something in the world outside ourselves that we can then share. We make connections to other human beings to the extent that we can agree on our shared cultural experiences. We make sense of ourselves only in relation to the narrative of our lives. And that narrative is as much written by others as it is by our own selves. We are shaped by the investment we make in others.

5 | Human beings alive today are anatomically and genetically almost identical to those living 200,000 years ago.* But because the soft tissue of the brain leaves behind no trace in the fossil record,† we do not know how our brains compare. Art appears to be our best evidence that we became truly modern about 40,000 years ago.

The so-called Lion Man is one of the oldest artworks yet found. Carbon dating tells us that it is around 40,000 years old. Carved out of a mammoth tusk, it was discovered in 1939, in fragments, in a cave in Germany. It seems it had been deliberately broken up – a difficult task to accomplish – but for what reason (perhaps ceremonial) can only be guessed at. The fragments – around two hundred of them – were archived and then forgotten about. About thirty years later they were rediscovered, and it was realised that they appeared to fit together. A first attempt at a reconstruction resulted in a body without a head. In 1997 additional pieces were located, and the sculpture we see today was revealed. It stands almost a foot high, a figure with the body of a man and the head of a lion.

Using fMRI, researchers have shown that when we are asked to think separately about a cat and then a man, the pre-frontal cortex is not involved. But when we are asked to imagine a cat-man the pre-frontal cortex hums into activity. It has been known for some time that the pre-frontal cortex is associated with abstract thinking. The appearance of abstract art around 40,000 years ago may be evidence that the pre-frontal cortex evolved around the same time.

Apart from tiny genetic differences that reveal themselves in variations in skin colour, height, hair texture, the ability to digest milk and

---

* But around 50,000 years ago the human population was reduced to a single group of people living in East Africa numbering no more than a few hundred. That this happened is not in doubt. Why it happened is the subject of much conjecture. Members of this small group left Africa and eventually became the seven billion of us who live on the earth today.

† Rarely, a mould of the inside of a fossilised brain case will reveal some of the details of the external surface of the brain.

so on – whatever differences there are between human beings born in the last 40,000 years appear to be differences of cultural heritage.

6 | On 18 December 1994 three cavers discovered by chance what is now known as the Chauvet cave. They found the remains of fires and animal bones. A bear skull seemed to have been deliberately placed on a rock, where it had remained for thousands of years. There were footprints still visible in the dust on the cave's floor, left by the last human visitors, before a landslide had sealed off the entrance. The cave is most famous for the coloured shapes of animals painted on its walls, exquisite drawings of aurochs,* horses and rhinos that appear three-dimensional and kinetic when seen by the light of flaming torches.

It has been said that creativity is a sign of a stable civilisation. If this is true, what are we to make of the fact that only time separates the paintings in the Chauvet cave – according to carbon dating around 32,000 years old – from those at Lascaux,† a 'mere' 18,000 years old? Visually and viscerally there is little to distinguish between the work of artists separated by a period (commonly referred to as the ice ages) six times as long as that which separates us from the days of Plato.

Yet among the cave paintings and artefacts that date from this long era there are works that are recognisably masterpieces, and others that are clearly amateur efforts. That we feel confident that we know which are which collapses the time between then and now. Aesthetics connects us with our ice-age ancestors.

Or so we think. Certain abstractly-shaped stones found at ice-age sites have only recently been recognised as representations of the female form. They suggest females only obliquely, sometimes in the extreme abstraction of no more than three or four triangles. Biometric evidence has been called on to confirm that these stones do indeed represent women. Many of them had been lying around in boxes in

---

* A kind of ox that became extinct in 1627. Modern cattle may be a sub-species.

† Discovered, also by chance, in September 1940 by Marcel Ravidat, then eighteen years old.

museum store rooms for decades, as if waiting for the (re)discovery of abstraction by twentieth-century artists like Picasso and Brancusi. Only now can we see the stones for what they are. Perhaps art has not developed at all in 40,000 years, only come full circle – not even full circle: what else might we not be seeing?

Across the ice-age world, semicircles, zigzags and other symbols crop up in cave paintings and scored onto artefacts, as if meant to represent something: perhaps flowing blood or water, social status, wounds, arrows; perhaps they are maps of terrain or of settlements; or symbols of something else altogether that we have not yet guessed at.

Why are the ice-age models of human beings nearly all of women? Why are they mostly exaggerated: large breasts, prominently outlined vulvas, large buttocks? Some archaeologists claim them as adolescent male fantasies, others associate them with childbirth. Some find them grotesque, some beautiful. Choose a story you like best.

Why are there so few representations of humans painted on the cave walls? Why, when the paintings of animals are so naturalistic, are there almost no naturalistic images of humans?

There seems to be a hierarchy among the animals, but we do not know why.

There is no human conflict depicted in cave art,* there are no depictions of landscapes, clouds, plants, the sun, moon or stars. Why not? We do not know.

Many of the images of animals are thickly outlined, the silhouette deeply and carefully etched. Nothing in nature has a line around it. As children, when we first learn to draw, we abstract the world by putting a line around what we take to be separate things. We transform the world into things represented by symbols.

There are thousands of surviving hand silhouettes at Chauvet, Lascaux and on rocks in Spain, China and Australia. They were

---

* Except at three separate sites, curious drawings of a figure with the limbs and torso of a man, pierced through with spear-like cuts.

produced by blowing a mouthful of paint as a fine spray around an outstretched hand. The image of a hand dipped in paint and impressed on rock would not have survived long, but a fine spray of paint adheres with extraordinary permanency. It is as if our ancestors were determined to reach out to us from the past, to raise hands of salute across the ages: do not forget what you once were. The hands are recognisably ours, except that in numerous cases fingers or parts of fingers are missing, perhaps lost to frostbite, perhaps to some prehistoric ritual whose significance is now lost to us.

Footprints surviving in the dust of the Chauvet cave are suggestive of dancing, as are the shapes of some surviving figurines. Remains of flutes have been found across the ice-age world.* What kind of music was played we will presumably never know.†

In recent times cave art has been linked to shamanism. The notion that the spirit world pervades the physical world has survived in aboriginal cultures on all continents. Was cave art produced as a result of chemical enhancement? It is known that the air in underground caves is often high in carbon dioxide. Scientific experiments have been carried out which show that elevated levels of carbon dioxide produce hallucinations specifically of animals and monsters. In shamanism this state is also created by music, pain, fasting, solitude and repetition.

Why do caves and cave-like passages haunt our dreams, and not memories of our tented life on the desert plains? On the threshold of the cave, squeezed into a narrow passageway, we find ourselves at the boundary between the body – a self carapaced in the stone-like armour of a skeleton – and our limitless spirit. We describe caves as being like cathedrals, but surely it is the other way round. We built cathedrals because they reminded us of caves. In the stations and hallways of

---

* Remains of ivory flutes found in caves at Geissenklösterle in Germany are at least 40,000 years old.

† In his film about the Chauvet cave, Werner Herzog wondered if any smells had survived from the ice age. He brought in a perfumier to sniff out molecules of the past. The perfumier could detect nothing.

underground railways, cathedral-like spaces knitted together and accessed by tunnels – a reminiscence of the Underworld – we go on shaping and building the subterranean dream-caves of our vestigial memories.

We ask certain questions of our alien ancestors, and other generations will undoubtedly ask different questions. Even if we never find definitive answers, the questions, and our partial answers, provide clues both to our nature as questioners and to the subject of our gaze.

We recognise the beauty of this prehistoric – what used to be called primitive – art, but we do not know what we are looking at. We do not know if we can see what they saw – what we once saw. We look back at ourselves as into a dimly silvered mirror and wonder who it is we see there.

Art makes things visible.
*Paul Klee (1879–1940), artist*

7 | In *The Selfish Gene* Richard Dawkins posited a complement to the gene called the meme. It was meant to account for the cultural evolution of human beings, the speeded-up rate of evolution that happened when humans got together and took themselves indoors.

Memes are units of cultural behaviour that are passed from one person to another and spread like measles, but instead of a virus it might be a fairy story that catches from generation to generation and across cultures. And so because of memes the best stories spread across the world. Memes unite to form memeplexes, which are meant to account for complex structures that take root in the world, whether in financial and legal institutions, in sport, the arts, or science. Some of these memeplexes are parasitical: spam emails, cults, alternative medicine, and religion. Winning parasite memeplexes like religion survive because of threats like hellfire. The problem, critics have pointed out, with this cultural parody of genetics is that there is no mechanism to tell us which are the good memes and which the parasitical ones. Bad genes are removed by death, but what removes bad memes? And if they are not bad, then what are they? It seems to be impossible to make any predictions about which cults will spread fastest. The meme theory is said to lack rigour. Stephen Jay Gould said memes are a meaningless metaphor. Unlike genes, memes have no actual physical existence, a curious property in a materialist model.

8 | Humans are assemblages of genes in an environment. Our conversations, the games we play, the meals we share, all the small ceremonies of life are constantly influencing, changing and shaping our brains, and influencing, changing and shaping how our genes get expressed.

9 | Evolutionists tend to concentrate on the genetic debt we owe to our deep ancestors, but we surely owe at least as great a debt to our recent ancestors. Denis Dutton, the founding editor of the web portal *Arts and Letters Daily*, says that our aesthetic preferences were forged in the 80,000 generations of the Pleistocene era, but as Raymond Tallis has pointed out, why privilege the Pleistocene over the five hundred generations since the first societies?*

---

* A lot has happened in the last 10,000 years. Of the numerous possible versions of human history that could be constructed, one might go like this: i By 7000 BC a fortified farming community at Jericho of some seventy-four acres. ii Traces of wine found on pottery at a Neolithic site (dating from around 5400 BC) in Iran are the earliest evidence of wine found so far. Traces of opium dating from 3,500 years ago have been detected on shards found in Cyprus. Less exotically, traces of cabbage have been identified on ancient British pots. (Perhaps things haven't changed so much after all.) iii The oldest known piece of creative writing dates from the third millennium BC: *The Epic of Gilgamesh*, a summary of legends from Babylonia, a state in the south of Mesopotamia. It is the story of the King of Uruk, and mentions many of the first city-states around which civilisation evolved: Ur, Eridu, Lagash and Nippur. It also contains the first account of a great flood, and the first account of a dream. In the Bible we are told that Abraham, the father of the Hebrew and Arab nations (the Israelites were descended from his son Isaac, and the Ishmaelites from his son Ishmael), travelled from Ur of the Chaldees. Chaldea was a region of Babylonia. Excavation of these ancient cities show us that they were sizeable: Eridu was twenty-nine acres, Ur twenty-five acres and Uruk 173 acres. iv The reign of the Pharaohs began in 3100 BC. Weather is a recurring theme of early history. In recorded histories the fate of Egypt is often

associated with times of drought and famine. The first settlement of Egypt was in the delta of the Nile, at a time when there were many more tributaries than there are today. The Egyptians' name for their own land was *Kemet*, which means black earth. This black earth is the alluvial deposit from the annual, thought to be miraculous, flooding by the Nile of the delta plain. The surrounding land was called *Deshret*, or red land, from which we derive the word desert. *Kemet*, with its magical properties of fertility, will become the word alchemy, and thence chemistry. v The history of early Greek civilisation from 1600 BC to 1100 BC is called the Mycenaen period. Mycenae was a city about ninety kilometres south-west of Athens. It was in this period that Helen of Troy eloped with Paris, the consequence of which is the subject of Homer's *Iliad*. Troy was named Ilion, hence Iliad, the story of Troy. Life expectancy at this time was thirty-five for men and thirty for women. vi The history of ancient Greece begins around 1100 BC. Some say ancient Greek history proper began with the first Olympic Games in 776 BC. vii In 1000 BC there were many communities along the western Atlantic coast of Europe peacefully trading with each other. It is surmised that trading first began because individual fishing communities met up with each other as they chased fish from feeding ground to feeding ground. There were stable fishing communities in this region for hundreds of years. It is to one of these seafaring tribes that Celtic history can be traced. viii There were Olmecs in Mexico from 1000 BC and Chavin in Peru from 900 BC. ix Greek history went through a period of five hundred years that is sometimes called the Greek dark ages. But by 600 BC Greek civilisation had become what Nietzsche described as 'the most accomplished, most beautiful, most universally envied of mankind'. Some historians claim that Greek philosophy started on 28 May 585 BC. It was on this date that Thales of Miletus (c.624–c.546 BC), the first of the pre-Socratic philosophers, correctly predicted a solar eclipse. Greek civilisation ended with the death of Alexander the Great in 323 BC. Egypt as the land of the Pharaohs comes to an end at the same time. x Taoism was founded by Lao-tzu (b.604 BC), and expounded in the 5,000 characters that constitute the *Tao Te Ching* (*The Way of Virtue*), characters that are familiar to every generation of Chinese since. Confucius, or Kung Fu-tzu (551–479 BC), claimed that the Taoists disliked him. Certainly he was more practical, more socially applied. What are we to make of the fact that there have been no holy wars in China? xi Around 400 BC Celts found their way to the Mediterranean coast, drawn there by the allure of figs, grapes, oil and wine. What we know of these people from accounts written by the Greeks, and later the Romans, is that they liked drinking and were said to be somewhat flashy: they wore gold ornaments, of a kind we would now describe as bling. Even at the time, what the Greeks had to say about the Celts was a generalisation. The Celts were known as the barbarians of western Europe, partly because their vanquishers, the Greeks, wanted to elevate their own accomplishment in subduing a violent and brave race, while at the same time ensuring that it was under-stood that, as victors, they were braver and more civilised. There are populations today

10 | When you come across a bear in the wild it is likely that your pupils will dilate, your heart beat faster, your intestines contract, your muscle tension change, and that you will sweat and produce extra adrenaline. The brain processes these changes into a feeling. In this case not a good one. Sometimes when you listen to the right kind of music your pupils dilate, your heart beats faster, your intestines contract, your muscle tension changes, you sweat and produce extra adrenaline. The brain processes these changes into a feeling. Our fear of bears can be explained from a survival point of view, as reflex reactions. When we

---

claiming to be Celtic in Ireland, the UK, France and Spain, and there are reasons to suppose that there is a connection between these peoples because of connections to be found in their spoken and written languages. The word Celt wasn't revived until the eighteenth century. The mythology is essentially modern, yet there is a Celtic history that supports the mythology in telling ways. xii How did a small, apparently undistinguished, town in the middle of Italy take over the world? If the Greeks seem to have been essentially philosophical and ambivalent, the Romans were always looking back to a Golden Age even during their own golden age, and looked forward to what they saw as a future of inevitable degeneration. Their prophecy was largely fulfilled, of course – perhaps self-fulfilled. The degeneration of Roman civilisation is mirrored by a decline in Roman writing, which by the third and fourth centuries AD had become vacuous; though even here so much is a matter of taste. Ovid, whom we now think of as one of Rome's greatest writers, was until fairly recently dismissed for his triviality. Rome's heyday was in the second century AD. An empire of sixty million humans was spread out across an area some twenty times the size of Britain. The lands circled the Mediterranean, known affectionately to Romans as *Mare nostrum*, our sea. There was trade between Rome and India. Spices, muslin, jewels and ivory were exchanged for gold. By the time of Rome's final decline the story shifts to that of its most significant outsider citizens, the Christians. xiii In India the period between the third and sixth centuries is known as the Golden Age, largely because of discoveries in astronomy, mathematics, religion and philosophy. xiv From the fifth to the tenth centuries: the so-called Dark Ages in Europe. xv Between AD 1000 and 1100 the population of China doubled because of expanded rice cultivation. xvi The first crusade 1095–99. The ninth crusade 1271–72. xvii In China in 1368 a peasant named Zu Yuanzhang overthrows the Mongols and founds the Ming dynasty. xviii In 1521 the Aztec empire falls. xix In Europe from the fourteenth to the seventeenth centuries, the Renaissance. xx In the early twentieth century Britain rules a fifth of the world's population, the largest empire in history. xxi In the mid to late twentieth century the world is dominated culturally and economically by the USA. xxii China?

listen to music we have coopted those reactions, but what is it in the airwaves that we are reacting to? We are reacting to a shared cultural heritage that has been encoded on the page and realised in a performance. We are responding to collective and singular human expressions of meaning that have meaning only for other humans. We are tempted to elevate the survival explanation because it comes first in an evolutionary story, just as we want to elevate the universe at the Big Bang over everything that came afterwards because it comes first in a cosmological story. A symphony, which comes at the end of this story, is physically insignificant compared to the Big Bang; all its significance is human-scale.

Musical behaviour such as chanting, singing, dancing and playing instruments is universal among humans and even societies that lack writing and complex organizations practise highly developed rituals based on forms of music.
*Jill Cook, Ice Age Art*

If science breaks us apart, art puts us back together.
*Jonah Lehrer*

Art is the nearest thing to life: it is a mode of amplifying experience.
*George Eliot*

Works of art have and need no justification but themselves. Art is a self-justifying activity in the same way (probably) that life is, asserting its own importance instinctually, and vindicating its importance not by where it is going but by the intensity with which it is.
*Brigid Brophy (1929–95), novelist and critic*

It is only our words bind us together and make us human.

*Montaigne*

11 | For the philosopher Jacques Derrida (1930–2004), language has meaning because it grants us the power to value ourselves as a community.

12 | Language is a reality between people. Like motion it is meaningless except as a relationship between bodies. Motion means nothing in a universe with only one thing in it. And so with language the existence of at least one other human being is necessary for it to become what it is, an agency of meaning. We know that there are other things in motion so long as we are willing to accept that there are things that are separate from each other. Language becomes a reality between people, and then, paradoxically, out of the assumption of separateness that gives language its force, we may find, once more, occasionally, in love and empathy, that the boundaries dissolve between the self and another.

13 | In later life Ludwig Wittgenstein* acknowledged that words do more than express facts. There are many ways – perhaps an infinite number of ways – in which words can be exchanged between humans meaningfully, each exchange understood in its own context as, say, a command, a promise or an emotion. Of course what we mean by commands, promises, etc. also has to be defined within some context. In this way meaning is nested, rather as Gödel's theorem nests mathematics within itself.

---

* Wittgenstein claimed that no one really understood what he was trying to say. In the 1920s, when logical positivism was at its most influential, he recited poetry at meetings of the Vienna Circle. Perhaps surprisingly for someone who wondered to what extent words mean things, Wittgenstein was garrulous. Paul Dirac said he was an 'awful fellow … never stopped talking'. Dirac himself was famously laconic: 'What would you say if I told you I was leaving?' his wife once asked him. 'I'd say goodbye,' he said.

There are, indeed, things that cannot be put into words ... They are what is mystical ... What we cannot speak about we must pass over in silence.
*Ludwig Wittgenstein in full Biblical style, Tractatus Logico-philosophicus*

The meaning of a word is its use in language.
*Wittgenstein, Philosophical Investigations*

14 | Words are discrete, reality flows. But we do not use words one at a time. There is conversation, and when conversation flows language becomes the integral of its parts and meaning emerges.

15 | Words that win a stay of execution or proclaim a sentence of death. Words that win love. Words that have the power to hurt. Words and their manifest manifesting powers.

16 | If talking really does cure, it does it not through direct physical means, molecule on molecule, but through the intermediary of meaning and understanding.

He never fully realised ... what a deep channel they cut; it never occurred to him that these words, uttered loudly, squarely and without any tinge of false embarrassment but boldly and unabashedly, do not just disappear without trace ... but sink deeply like pearls into the silt of social intercourse and always end up finding a home in an oyster shell.
*Ivan Goncharov (1812–91), Oblomov*

Many people hesitate to put in a good word for the good and redden with embarrassment at the thought, but think nothing of throwing in some casual, flippant remark, little suspecting that unfortunately, such remarks, too, do not just disappear into thin air but leave behind their own long-lived, sometimes indelible traces of evil.

*Ibid.*

Narrative is the basic modality in which human mind functions.
*Thomas Berry, cultural historian*

Our cortex makes up stories about the world and softly hums them to us to keep us from getting scared at night.
*Leif Finkel, a bio-engineer at the University of Pennsylvania*

17 | There is no escape, except by dying (and temporarily from drugs and suchlike), from our own existence. Life may be meaningless, but even the most unplanned or aimless life becomes a biography over time; a story that seems to have some quality of inevitability about it. Things have a way of seeming inevitable whichever way they turn out. The universe, too, is inescapable, and – like life – has a look of inevitability about it when seen in retrospect.

It is far from clear whether or not it is possible to construct an account of the universe free of our perspective as storytellers. It is our human nature to assume that everything means something, even when we have no idea what that something is. We live as if there is meaning.

That we are not the end point of creation seems clear enough, though we are the first, it seems, who can tell the story.* Our telling of the story may be crucial to the tale itself, a philosophical conundrum out of which there appears to be no escape. We are in the universe. There is no outside from which to tell the story, only outsides that we create in imagination.

---

* We may also be the first who are close to having the power to destroy all life on earth, or even the power to destroy the entire universe (which, were we to exercise such power, would, ironically, prove us to have been in a privileged position in the universe after all).

As part of its methodology, and even though we are clearly the ones telling the story, science tries to minimise the role humans play in the story of an evolving universe.

The compulsion to construct narrative is the default mode of being human. Hindsight as foresight makes no sense, said W.H. Auden, but we have no choice as scientists but to tell the story from here, and rerun it as if it ran forward all along. That is the price we must pay for having a cosmology at all.

18 | i We are not always in control of the stories our bodies and our brains are telling us. A person with Capgras Syndrome loses her emotional ability to identify those around her, even though she retains her capacity to recognise them intellectually. If a close family member enters the room, a large gap opens up between the knowledge that here, say, is her mother, and the emotional feeling that this is a stranger. In order to bridge the gap a story is constructed. Typically, the story is that this person must be an impostor, someone who looks exactly like her mother but is in fact an alien or a zombie. Capgras Syndrome appears to be proof that emotion trumps intellect. We feel our way through the world. Even our most intellectual thinking is a kind of processed emotional response to the world. In Cotard's Syndrome all the emotional centres of the brain become disconnected from the senses, and the subject concludes that he must be dead. In Anton's Syndrome the subject is blind but won't admit it, fumbles around the world as if she can see. Those with blindsight claim that they cannot see, even though they can point to a named object when asked to. Patients with anosognosia believe that everything is well even when they have serious injuries. There are those who cannot make abstractions* and are

---

* By abstracting we make connections between things, and so prune the world down to a place we can negotiate and about which we can tell stories. The fly we see from this angle we recognise as the same fly seen moments later from a different angle. The risk is that sometimes, perhaps often, we are wrong; that it really is a different fly, or even something we have mistaken for a fly.

overwhelmed by every fact; everything elicits a connection to something else. For some the body is at war with itself, one hand doing up shirt buttons as the other hand undoes them.

ii In brain studies it is often the abnormalities that throw light on the complexities of what it is to be normal. Case studies of neurological disorders are important because they reduce the human experience to something against which 'normal' experience can be measured and more fully described, but we can become so mesmerised by abnormal functioning that we discount what is normal. We are inclined to elevate abnormal skills: those who have no power to forget, or who can manipulate long numbers together in their heads. We see that certain functions can be heightened when brains go wrong, but forget that it is at the expense of something missing or broken. We tend to downgrade the normal merely because it is usual, but we might wonder why nature goes to so much trouble to make us as similar as we are.

> In this scherzarade of one's thousand one nightinesses that sword
> of certainty which would identified the body never falls.
> *James Joyce (1882–1941), Finnegans Wake*

> Humans are the animals that believe the stories they tell about
> themselves.
> *Mark Rowlands, The Philosopher and the Wolf*

19 | The American newscaster Dan Rather once said of the Chernobyl disaster: 'If it weren't for the wind, no one would know this story.' This is partly true; what he omits to add is that without human beings to tell the story there never are stories.

> All our inventions are true, you can be sure of that. Poetry is as
> exact a science as geometry.
> *Gustave Flaubert (1821–80), novelist, in a letter*

20 | In AD 350 Sappho first applies the epithet 'silver' to the moon.

Everywhere I go, I find a poet has gone before me.
*Sigmund Freud (1856–1939), founding father of psychoanalysis*

It is quite possible – overwhelmingly probable, one might guess – that we will always learn more about human life and personality from novels than from scientific psychology.
*Noam Chomsky, linguist and philosopher*

Shall we for ever make new books, as apothecaries make new mixtures, by pouring only out of one vessel into another?
*Laurence Sterne, Tristram Shandy*

It was books that taught me that the things that tormented me most were the very things that connected me with all the people who were alive, or who had ever lived.
*James Baldwin (1924–87), writer*

A novel isn't a piece of ethics or sociology. It is a release of certain energies and a dramatisation of how these energies might be controlled and given shape.
*Colm Tóibín, novelist*

21 | Every book owes something, perhaps almost everything, to all books that have gone before, is part of one collective book that the human race has been writing since *The Epic of Gilgamesh*. The influence of this collective book stretches beyond its writers and readers.

22 | The evolutionary epic is probably the best myth we will ever have, wrote the biologist E.O. Wilson. But it is also the first account of creation to spurn both meaning and metaphor.

23 | The ancient Greeks had two ways of telling the story of creation: through *logos* and through *mythos*.* They mixed the two ways of

---

\* We might, say, take the bull as a mythic clue that threads across the world's civilisations. i The story begins with Zeus, the king of the gods and ruler of Mount Olympus. Zeus had many offspring, including Perseus, Helen, Hercules, Hermes, Artemis and Persephone. By Mnemosyne he fathered the nine muses (Out of Power and Memory comes Art). By Europa, whom he had ravished in the form of a bull, he fathered Minos, the mythical king of Crete. The story goes that the god Poseidon sent Minos a pure-white bull that Minos was meant to offer in sacrifice, but Minos was so taken by the bull that he decided to keep it. As punishment, Poseidon caused Minos's wife, Pasiphaë, to fall in love with the bull. The Minotaur was the issue, a wild creature half man and half bull. In an attempt to contain its violence, the man-bull was housed at the heart of a labyrinth, built for Minos by Daedalus and his son Icarus. To appease the Minotaur's wrath, every seven years seven young Athenians of each sex were offered to the man-beast in sacrifice. Before the third cycle of sacrifice, Theseus, a young Athenian, Poseidon's son, offered to kill the Minotaur, and though he did not doubt his own strength, he did wonder how he would escape the labyrinth once he had accomplished his task. Ariadne, Minos's daughter, who had fallen in love with Theseus, gave him a ball of red woollen thread, or 'clew'. Ariadne gave Theseus a clue and Theseus followed it. He secured the thread as he entered the maze, and unwound the clew as he found his way to the heart of the labyrinth, where he slew the Minotaur. He found his way out again by rewinding the thread of the clew. The Minotaur had become a myth even by the time of Homer. In the nineteenth book of the *Odyssey* Homer writes of a fertile island – a reference to Crete, probably – a place where there are countless people and ninety cities. In the *Iliad* it has grown to a hundred cities. We do not know when Homer lived, or whether he was a tradition rather than a single writer, but even the Trojan War that is the subject of the *Iliad* happened over a century after the Minoan civilisation had disappeared. Plato associates the lost Minoan culture (long lost to history even by his day, the fourth century BC) with the lost city state of Atlantis. Physical evidence of the Minoan culture was discovered only in the late nineteenth century. Until that time, it wasn't known for certain that there had even been such a culture. All that changed in 1878 when, at Knossos on the island of Crete, the remains of the palace of Minos were discovered by a Cretan merchant named, appropriately enough, Minos Kalokairinos. The German excavator Heinrich Schliemann (1822–90), driven by a belief in the historical reality of places mentioned in Homer, continued to explore the site, but the full significance of the remains only became apparent after the site was acquired by the British archaeologist Sir Arthur Evans (1851–1941). The many discoveries that were made there are recorded in his monumental four-volume work *The Palace of Minos* (1921–35). The Minotaur's labyrinth has been placed at the north-west corner of the palace. Among the many important works of art discovered there

storytelling together. In trying to understand the universe and the human condition *logos* can take us only so far.

---

are wall paintings of young men and women leaping bulls. It has been suggested that European culture began right here, in the Minoan culture that is thought to have flourished between 2700 and 1450 BC. Minoan art and pottery has also been found at ancient Egyptian sites, and Egyptian pottery at Minoan sites. As recently as 1987, fragments of Minoan wall paintings were found in Egypt that depict bull-leapers like those seen in the palace at Knossos. The finds cast some doubt on whether Minoan culture even comes from Crete, as has been supposed. It may have arisen in Egypt. It is possible that the Minotaur's labyrinth refers to an Egyptian labyrinth sacred to the sun. Or that Minoan civilisation may have arisen somewhere else altogether. The rediscovery of this lost culture also retrieved a lost chapter in the history of world religion. Despite the later Greek mythology, the Minoans did not worship a bull god; so far as we can tell, their gods were all goddesses. There was a goddess of fertility, a goddess who protected cities, another who protected the household, another the underworld, and so on. The significance of the Minoan frescoes of bull leaping are as Greek to us today as they were to the Greeks, but it is clear that they portray a dangerous yet graceful and athletic interaction between youths and bulls. It is impossible not to imagine that some echo of this ancient ritual survives in the graceful bullfights found in Hispanic cultures today. What is missing in the Minoan culture is any sense of barbarism. The relationship between the bulls and the acrobatic Minoan youths seems to be one of harmonious enjoinment. ii In the caves at Chauvet, Lascaux and elsewhere are exquisite, detailed paintings of aurochs. In *The Epic of Gilgamesh* (from the third millennium BC) Gugalana is named as the Bull of Heaven. The horns represent the crescents of the moon. In Ancient Egypt the bull was worshipped as Apis. Many Egyptian rulers held the title 'mighty bull' and 'bull of Horus'. The Egyptian word for bull was *ka*, which is phonetically identical to a word that describes the king's divine double, a sort of bull doppelgänger or daemon. In the Marduk civilisation of the Indus Valley (around 1800 BC) we find the bull of Utu. In Hinduism Shiva's mount is a bull named Nandi. In the book of Exodus Aaron's Golden Calf was worshipped by the Hebrews. An ox traditionally (though there is no mention of it in the New Testament) witnesses the birth of Christ. In Greek mythology Dionysus is slain in the form of a bull and eaten by the Titans. Alexander the Great named his horse Bucephalus, literally 'ox head'. In Rome the god Mithras is often depicted slaughtering a bull. In the first century AD Pliny the Elder wrote of a Celtic druidical ritual in which white bulls swathed in mistletoe were sacrificed. iii 'It is a story about a Cock and a Bull – and the best of its kind that ever I heard!' Laurence Sterne, *Tristram Shandy*.

# On the relationship between human beings and nature

A change in the weather is sufficient to recreate the world and ourselves.

*Marcel Proust*

1 | About 70,000 years ago the world entered a glacial period, which reached a peak of coldness about 21,000 years ago. By 10,000 BC the climate was as warm as it is today, and the warmest it had been for hundreds of thousands of years. Tools and a change in the weather enabled us to farm. Humankind selected just a few plants and animals, and began to alter the world forever. Once-nomadic humans now settled down as small communities. Some of these communities grew in size. For the first time, a species moved outside its own local ecosystem. For the first time, a species began deliberately to control the environment. Human history has been a process – powered by technological progress – of moving out of nature and indoors.

2 | No one can control the weather, but collectively humans have managed to change it. For the last 12,000 years we have been living through an interglacial period – an unusual and brief period of clemency. During this time the earth's temperature has increased. Because of human activity the temperature is now increasing faster than it otherwise would.

There is a direct correlation between levels of carbon dioxide ($CO_2$) in the atmosphere and the temperature of the earth. For 12,000 years the concentration of $CO_2$ stayed constant at about 260 parts per million. In the 1800s, as the industrial revolution spread widely, it began to climb; by the end of the century to around three hundred parts per million. It reached 346 parts per million by 1985, 378 by 2005, 387 by 2009, and four hundred by 2012.

The Intergovernmental Panel on Climate Change (IPCC), a body set up by the United Nations in 1988 to investigate climate change caused by human activity, warns us that at some time in this century the surface temperature of the earth will have increased by at least 1.1 degrees Celsius, and perhaps by as much as 6.4 degrees. An increase of a single degree will mean that the earth is hotter than it has been at any time in the last two million years. The earth hasn't been as hot as it is today for about a million years. Even if humans manage to restabilise the levels of greenhouse gases in the earth's atmosphere, the earth will continue to increase in temperature for the next thousand years. Within a few hundred years or less, the planet may no longer be able to support multi-cellular life like trees and humans. The earth will be returned to the rule of bacteria.

3 | In the 1780s, one of the first balloonists, moving silently, high above the ground, saw the earth from a new perspective, 'as a giant organism, mysteriously patterned and unfolding, like a living creature'. The earth has been thought of as a living organism for millennia, but it was James Lovelock in the twentieth century who first gave this belief scientific credibility. 'Gaia' is the name he gave to the now widely accepted hypothesis that there is an intimate connection between the earth's living and non-living processes. Gaia was an ancient Greek personification, literally grandmother of the earth.* When Lovelock first presented his ideas in the 1960s, they were largely ignored. There was a time when he was the only person in the world measuring the presence of CFCs

---

* The name was suggested to Lovelock by his friend and neighbour the novelist William Golding one day when they found themselves walking together to the village post office.

in the atmosphere. Lovelock predicts that by the end of this century Gaia will have established a new equilibrium, and billions of humans will have died.*

> Nature wastes a thousand seeds, experiments lightly with whole civilisations.
> W.N.P. Barbellion

4 | Of the perhaps fifty billion species that have ever existed on earth, evolution has done for well over 99 per cent of them.

5 | We are in the midst of a massive extinction, larger than the last mass extinction of sixty-five million years ago. This current extinction is greater than it would otherwise be because of the influence of humans.

Conservatively estimated, there are ten million species living on the earth today. We have identified and named about two million of them. Somewhere in the region of 30,000 species a year – microbes, fungi, plants and animals – are becoming extinct, many, presumably without our awareness that they ever existed.

> I can never get used ... to the opulence of nature and the squalor of human life.
> Vladimir Nabokov (1899–1977), Ada or Ardor

6 | If humans disappeared, life would flourish. If insects disappeared, all life would end within fifty years. A calculation made by Jonas Salk (1914–95), American virologist.

7 | Scientists and naturalists respond in radically different ways to Gaia. Lovelock says we must take control of the planet and wrest it back

---

* I imagine a kind of *New Yorker* cartoon: two businessmen surveying the wasted earth, one saying to the other: 'Don't worry, it's just a market correction.'

into a state that will continue to support human life. Only further technological intervention can repair the damage older technology has wrought. He has suggested, for example, that we add sulphur to aviation fuel, in order to create sulphuric acid droplets in the stratosphere to reflect sunlight back out to space.

> When we understand how the natural systems regulating the climate react to our technologies, and we begin to operate our technologies and economies so that they work in harmony with the climate, we will have transformed the divide between the natural and the artificial on a planetary scale.
> *Lee Smolin*

8 | The thin blue line of our atmosphere may be the only such line in the universe.

9 | The oceans move heat around the earth, the mountains create weather systems, volcanoes pump gases into the atmosphere, plants photosynthesise, animals breathe – all part of a balanced feedback system which extends out beyond the earth to include the moon and the sun. After billions of years of interaction between these living and inanimate structures, the conditions for life on earth are finely tuned.

We do not know what an ocean is. For now we know it as a simple, predictable, mechanical system, even though we also know that it must be much more complex, inseparable from the sky, inseparable from the workings of the whole earth.

> Every year the rivers bear thousands of tons of mercury, cadmium and lead, and mountains of fertiliser and pesticides, out into the North Sea.
> *W.G. Sebald (1944–2001), The Rings of Saturn*

10 | In May 2007, 250 miles north-west of the Galapagos Islands, fifty tonnes of ground haematite was released into the ocean, an experiment

to see if levels of phytoplankton could be boosted as a way of trapping $CO_2$ from the atmosphere. To work, the phytoplankton needs to sink to the bottom of the ocean when it dies, and stay there. It is not clear that this is what happened.

11 | Strontium 90 has been detected in the milk of nursing women living near nuclear test sites. The insecticide DDT, which was never used in Antarctica, has been found in the fat of penguins. Dutch elm disease was caused by a fungus from Asia, and accidentally spread across the world on imported timber. When it was introduced into Australia from Hawaii the cane toad didn't do what it was supposed to do,* and has since grown stronger. GM crops may encourage super-resistant weeds. Myxomatosis in rabbits. Canine distemper in lions. Thalidomide. Warfarin and superwarfarin.

12 | CFCs have improved the quality of our lives indoors, but we now know that they suppress biological life, increase rates of cancer, damage the immune system, cause cataracts and harm the environment.

We have involved ourselves in a colossal muddle, having blun-dered in the control of a delicate machine, the working of which we do not understand.
*John Maynard Keynes (1883–1946), economist; here writing about the study of economics, but he might as well be describing any complex system that we do not understand, like the weather, or human beings*

---

\* It was meant to eradicate a certain kind of beetle that was devastating sugarcane plantations. The beetles, however, live at the top of the cane, and the toads are not good at climbing. The toads spread widely across north-eastern Australia, and caused the decline of a number of native reptile species.

The world is disgracefully managed, one hardly knows to whom
to complain.
*Ronald Firbank (1886–1926), novelist*

13 | Humans are the only species found everywhere on the planet.

14 | For maybe 40,000 years the world's human population stayed
constant at around a million, rising to ten million by 6000 BC, one
hundred million by 500 BC, two hundred million by AD 1, three
hundred million by AD 1000,* two billion by 1927, three billion by
1960, four billion by 1974, five billion by 1987, six billion by 1999, seven
billion by 2011, perhaps eight billion by 2025 and nine billion by
2045–50.

15 | 250,000 people are added to the world's population every day.

16 | The population of Rwanda rose from about one and a half million
in the mid-1930s to over seven million by the late 1980s. In 1994,
800,000 people were killed there in one hundred days. The cause of the
genocide has partly been attributed to the country's population density,
which is among the highest in Africa.

17 | The overpopulation of Easter Island led to a genocidal conflict
between the 'long ears' and the 'short ears' that did for that
civilisation.

18 | Kassites and Babylonians, Persians and Greeks, Spartans and
Athenians, Macedonians and Persians, Romans and Carthaginians,
Romans and Gauls ... Danes and English, Normans and English ...
Crusaders and Cathars, Egyptians and Crusaders, French and English,

---

* The population of Europe halved during the sixth to eighth centuries AD. The Black
Death reduced the world population from 450 million to 350 million. The world's population didn't return to its 1340 level for two hundred years.

Scots and Norwegians, Spanish and Aztecs ... Chinese and Tibetans, Siamese and Burmese, Spanish and Portuguese ... Brazilians and Argentines, Spanish and Cubans, British and Zulus, Japanese and Chinese ... the world at war against itself, civil war ... Christians and Muslims, Hutu and Tutsi, Catholics and Protestants, MacDonalds and Campbells, Sunni and Shia ...

> Yet live in hatred, enmitie, and strife
> Among themselves, and levie cruel warres,
> Wasting the Earth, each other to destroy.
> *John Milton, Paradise Lost*

19 | When the world's first atomic bomb was exploded on 16 July 1945 near Alamogordo, New Mexico, no one knew exactly what would happen. At least one scientist feared a chain reaction would be set off that would destroy the earth.

20 | 6 August 1945 Hiroshima. 9 August 1945 Nagasaki. The bombs that fell on the two Japanese cities killed around 150,000 humans. Many were atomised, leaving behind nothing more than a shadow.

> The Japanese had, in fact, already sued for peace. The atomic bomb played no decisive part from a purely military point of view, in the defeat of Japan.
> *Fleet Admiral Chester W. Nimitz (1885–1966), Commander-in-Chief of the US Pacific Fleet*

> Now with the release of atomic energy, man's ability to destroy himself is nearly complete.
> *Henry Stimson (1867–1950), United States Secretary of War in 1947*

21 | On 25 July 1946 the Baker atomic bomb exploded over the Marshall Islands in the Pacific. One of the islands was completely vaporised. Bikini atoll is still uninhabited. The test resulted in the first instance of

concentrated radioactive fallout. The chemist Glenn Seaborg called it 'the world's first nuclear disaster'.

22 | 31 October 1952, the first successful test of a hydrogen bomb. It was named 'Mike'. The small island of Elugelab was vaporised. Waves twenty feet high followed the explosion and stripped surrounding islands of vegetation. Radioactive coral was blown into the air, falling on ships thirty-five miles away.

> We don't know how the third world war will be fought, but I can tell you what they will use in the fourth – rocks.
> *Albert Einstein*

23 | The greatest threat to the world is unchecked religious fundamentalism. The greatest threat to the world is unchecked scientific fundamentalism.

> And God blessed them, and God said unto them, Be fruitful, and multiply, and replenish the earth, and subdue it: and have dominion over the fish of the sea, and over the fowl of the air, and over every living thing that moveth upon the earth.
> *Genesis, King James Version*

> Men ... Ransack'd the Center, and with impious hands
> Rifl'd the bowels of their mother Earth
> For treasures better hid.
> *John Milton, Paradise Lost*

> [Then, we] saw the world as an inexhaustible Bagdad Bazaar. Now we see it is exhaustible, and are grimly determined to exhaust it as soon as may be.
> *W.N.P. Barbellion*

[Man is] master and possessor of Nature. Which aim is not only to be desired for the invention of an infinity of devices by which we might enjoy, without any effort, the fruits of the earth and all its commodities, but also principally the possession of health, which is undoubtedly the first good, and the foundation of all other goods of this life.
*René Descartes*

Painters woo nature, scientists violate her.
*Jean Renoir*

24 | The pile of shit outside your castle wall was an emblem of wealth and hospitality. Look how well I entertain my guests! For centuries we have been inside the castle jettisoning our waste over the walls. The world outside has been our dungheap. But now the shit begins to seep back in: the castle walls are breached, and the inside becomes the outside.

Yeah, we are playing God, and it's a good thing. We play God all the time, starting with, you know, agriculture. We try to change the world, including forms of life, in ways that are beneficial. And it's important that we do so, because we've been able to prosper and flourish as a result of it.
*Mark Bedau, American philosopher working in AI*

We don't do it because it is useful, we do it because it is amusing.
*The biologist Eric Kandel, quoting an unnamed scientist*

All men naturally desire knowledge.
*The opening words of Aristotle's Metaphysics*

25 | Without science there would be no such thing as progress.

26 | Between 1700 and 1900 average life expectancy in Britain rose from seventeen to fifty-two.

27 | The farming methods of the 1940s could not support the world's current population.

28 | The introduction of fertilisers in the 1950s saw the number of malnourished people in the world fall by 20 per cent.

29 | In 1970, 37 per cent of the developing world was undernourished, now it is 16 per cent.

30 | The world produces enough food to support the world's population.

31 | Genetic modification has made it possible for rice to store beta-carotene, normally found only in rice leaves. A source of vitamin A has been made available to millions who would otherwise not get it.

32 | Smallpox was eradicated by international cooperation. It killed three hundred million people even in the twentieth century. By 1978 it had been eradicated worldwide.*

33 | Living pesticides are being used to destroy mosquitoes, with the aim of eradicating malaria. Sterile male mosquitoes are being bred which dominate the population and so wipe out the next generation. The technique has the benefit that it targets only a single species.

---

* Ali Maow Maalin, a hospital cook in Somalia, contracted the disease in 1977, the last known natural case. Janet Parker, a smallpox researcher in England, was exposed to the virus in the laboratory where she worked, and died on 11 September 1978. After her death all laboratory holdings were destroyed, except for some still held in Russia and the United States. A 2010 WHO investigation found that no useful purpose is being served by keeping the virus active.

34 | i When the American conservationist Dave Foreman was told by a US senator not to be emotional or he would lose his credibility, he said: 'But damn it, I am emotional. I'm an animal and proud of it. Descartes was wrong when he said, "I think therefore I am." Our consciousness, our being, is not all up here in the skullbox. It's our whole body we think with, and it goes beyond that. David Brower tells us that you can't take a California condor out of the wild and put it in the LA Zoo and still have a condor, because the being of a condor does not end at those black feathers at the tips of the wings. It's the rising thermals over the Coast Range. It's the rocky crag where she lays her egg. It's the carrion she feeds on. The condor is place … and we are place too.' A condor in a cage is little more than a lump of meat stuck with feathers.

> The condor, along with the frogs and salamanders that are vanishing, is a constant reminder that I am not the center of it all.
> *Paul Shepard, naturalist*

ii Five hundred years ago, the California condor could be found all the way down the American west coast and across the states of the southwest, but during the last century its population began to fall dramatically. By 1987 every California condor in existence could be accounted for, and there were just twenty-two of them. In a risky strategy, all twenty-two were captured and kept at Los Angeles Zoo, where they were successfully bred. A few were released back into the wild in 1991. In 2007 a California condor laid an egg in Mexico for the first time since 1930. Today the world population of the California condor has risen to 349, about half that number living free. Saving the condor has cost tens of millions of dollars, making it the most expensive species conservation project in history.

Is this a moving story of humans making some kind of reparation for the damage inflicted on nature by other humans, or another story of humans interfering with nature? Certainly no one doubts that the decline in the condor population was due to human influence. In the mistaken belief that they kill young cattle, ranchers shot condors. They also killed them indirectly. Other birds shot by ranchers using lead shot, and left to rot where they fell, became carrion for the condors. Because their stomach acid is strong enough to dissolve lead, many condors have died of poisoning. The California condor has a wingspan of between eight and ten feet. Large numbers of them have met their end by flying into electric power lines. And then there are the usual reasons why species are declining across the globe, because of habitat destruction and climate change.

Some naturalists were alarmed by the seeming recklessness of the strategy that was adopted to save the condor. Animals evolve complex habits, not all of them genetically determined. By raising condors in captivity, breeders were forced to interfere with the condors' usual and natural habits. Condors rear one chick at a time, but if the first egg is removed the female can be encouraged to lay a second. In order to speed up the captive breeding process, it was decided that the females should be encouraged to lay two eggs. The first-laid egg was hatched artificially, leaving that chick's parents free to bring up a second chick from an egg hatched by more conventional means (though still hardly under natural conditions). The first-born chick was brought up by human keepers using hand puppets as stand-ins for adult birds. Later, using a kind of aversion therapy, all the birds were trained to avoid flying into power lines. Effectively, the condors were partly domesticated. Humans have taught them some new survival techniques, but what has human interference removed? It is impossible to know in what ways the 'wild' California condors of 2010 differ from those few remaining condors before they were captured in 1987.

35 | Humans have been blurring the boundaries between the wild and the domestic for centuries. Joseph Hooker's *The Rhododendrons of Sikkim-Himalaya* was a sensation when it was first published in 1849–51. It ignited a craze for rhododendron-growing across Britain that has permanently changed the landscape.

When sheep were slaughtered en masse during the UK foot-and-mouth epidemic of 2001, naturalists were concerned that subsequent generations of sheep would lose a culturally-acquired homing instinct known as hefting. A hefted sheep lives on a hill without fencing, but never strays beyond a certain region. Large areas of hillside might be divided up into many hefts, containing a large number of sheep, but each sheep or lamb knows which heft it belongs to. Hefting is an ability that has co-evolved between humans and sheep over more than a millennium. Herdwick sheep, for example, probably came over with Norse settlers in the tenth century. The word heft comes from the Old Norse *hefda*, meaning to acquire by right or prescription. Hill farmers, when they acquired new land, would train their sheep to stay within a particular area without being herded, the area being known as the sheep's heft. Hefted sheep passed on that information to their lambs, and so on down generations of sheep for centuries. By removing entire populations of sheep from a particular heft, that cultural knowledge is in danger of being lost forever. Few farmers know any longer how to heft a sheep (hill farms themselves having often been passed down unaltered for many generations). And in any case, it takes years to heft a sheep from scratch. Without hillside sheep, the landscape will change, bogs and marshes will spread, and gorse and juniper will take over.

Thousands of years of farming has created a very particular kind of European pastoral scene. Much of the English countryside has been tamed and integrated into our history and culture. Wherever we come from, we cannot escape our environment. We are forced to have some kind of relationship to the environment we inhabit, on the outside (and 'other'), or on the inside (and 'civilised').

To learn to live with our planet, we have to rid ourselves of the vestiges of this old yearning for elevation from it ... We need to see everything in nature, including ourselves and our technologies, as time-bound and part of a larger, ever-evolving system.
*Lee Smolin*

Knowledge is power was the belief behind all the activity of the students of the world – but it was knowledge of the infinitely divisible atoms of that world. Out of the endless breaking up and recombining of elements, the numberless taking-to-pieces of things, the examination and the observation – out of all that knowledge people had learned how to move faster, how to transmit electrical force, and how to be more physically comfortable. The practice of infinite divisibility had produced an enormous number of devices. Indeed, the whole world was cluttered up with devices. Knowledge is Power – and with the power they gained, they learned to make electricity do their living for them. The more singleness, separateness, and indivisibility became the habit of our development (so that everywhere everybody was breaking away from old patterns of social and family life), the more ways there were of escaping mechanically. Actually, the conquest of machinery was to promote the separation of the individual from the mass; and the by-product of scientific conquest was to become the elaborate, unhappy, modern man, cut off from his source, powerful in mechanisms, but the living sacrifice of his scientific knowledge.
*Mabel Dodge Luhan, Edge of Taos Desert*

36 | Tools increased our sense of separation from the world. The separation of tools and self may become less apparent in the future as tools become integrated into flesh. Tools will become seamless extensions of the body rather than separate things that we can hold or touch. Perhaps then we will feel (again?) the continuity between the self and its environment.

# On the relationship between human beings and other animals

How incandescently, how incestuously ... art and science meet in an insect.
*Vladimir Nabokov, Ada or Ardor*

Little Fly. Thy summer's play
My thoughtless hand
Has brush'd away.
Am not I
A fly like thee?
Or art not thou
A man like me?
*William Blake, 'The Fly'*

My magic well.
*The ethnologist Karl von Frisch (1886–1982), of his bees*

1 | When we believe that animals are merely things, we are liable to treat them as things. When we begin to see that even flies are like miniature models of human beings, our circle of empathy has the potential to grow. A reductive investigation may start out by looking for differences between humans and animals – there must always be aliens of some kind if science is to make progress – but progress itself is often the discovery of what unites rather than divides. And always,

the ocean of what we don't know is behind us if we care to turn and gaze on it.

2 | When the biologist E.O. Wilson was asked by a friend what to do about the ants that had invaded his kitchen, Wilson said 'Watch where you step.'

3 | In 1903, Topsy the elephant was publicly electrocuted by Thomas Edison as a display of the effectiveness of electricity, in particular of direct over alternating current. The event can still be seen on film. Alternating current won out.

4 | On 6 December 1834 the *Beagle* anchored at the island of San Pedro. In his Journal of Researches for that day, Darwin wrote: 'A fox (*Canis fulvipes*), of a kind said to be peculiar to the island, and very rare in it, and which is a new species, was sitting on the rocks. He was so intently absorbed in watching the work of the officers, that I was able, by quietly walking up behind, to knock him on the head with my geological hammer.' The species is today known as Darwin's fox, and has been reclassified as *Lycalopex fulvipes*. It is critically endangered.

5 | In the so-called shuttle box experiment conducted at Harvard in the 1950s by psychologists R.L. Solomon, L.J. Kamin and L.C. Wynne, a dog is kept in a metal cage made of two compartments. An intense electric current is passed through the floor of one compartment. The dog instinctively leaps to the other. A single experiment might consist of hundreds of shocks being administered. After ten days or so of this the dog ceases to resist the shocks. The experiment turned out to advance scientific understanding not at all, and is now discredited.

6 | The neuroscientist Cori Bargmann used to work on mice, but she would cry every time she had to do anything invasive to them. She realised she would have to find a different animal to investigate. She

chose the worm, and read everything she could lay her hands on about them, including back copies of *The Worm Breeder's Gazette*.

He tied a slipknot in a wire to make a noose and looped the wire over one of the victim's legs. He cinched it tight, then suspended the hapless body over a pool of salt water. The victim waved and kicked his leg. Whenever the wire touched the water the body received an electric shock. Eventually the victim learned to keep his leg out of the water.

*(The victim is a fly. I have slightly anthropomorphised this description of an early experiment which showed that flies can learn.)*

7 | These days every laboratory in America has an animal welfare committee. 'We treat mice as if they are staying at a hotel,' says my friend Georgia. Her work led to the introduction – since 1982 – of taurine in baby food, with the result that the disease dilated cardiomyopathy no longer exists. Her research required her to kill kittens. She still has nightmares about it. 'You've no idea how hard it is. I had to cut myself off from my immediate feelings and focus on the greater good.' And though of course she is proud of her accomplishment at ridding the world of a childhood disease, it is important to her that her work also led to the introduction of taurine in cat food, so ending the disease in cats as well as in humans.

8 | Aged ten, Tristram Shandy sees his uncle Toby gently capture a tormenting fly. 'I'll not hurt a hair of thy head: – Go, says he … go poor Devil, get thee gone, why should I hurt thee? – This world surely is wide enough to hold both thee and me.'

The action, says Tristram, 'instantly set my whole frame into one vibration of most pleasurable sensation … the lesson of universal goodwill then taught and inspired by my uncle Toby, has never since been worn out of mind … I often think that I owe one half of my philanthropy to that one accidental impression.'

9 | We might use consciousness as a guide to which animals to include in our circle of empathy, if we could agree what consciousness is. Learning seems to be some kind of test of consciousness, and so we might begin by separating out those animals that can learn from those that are 'automata'.

10 | For Descartes, only humans have consciousness. Since his time the line has been drawn in different places. V.S. Ramachandran thinks that even late arrivals in evolution, like cats, do not feel pain in the way that human beings do. They are not conscious of it, he says: they react to it instinctively. The neuroscientist Christof Koch, on the other hand, thinks that we can safely grant consciousness to anything that can be taught to do something that isn't inborn. Understanding that flies and bees can be taught has made him non-violent even towards insects. In the early days of behavioural investigation of the fly, populations were investigated en masse; these days it is not unusual for individual flies to be scrutinised. We treat many animals as if they are all the same, until we look closely; and then we see the differences.

11 | Even among round worms it is possible to identify those that are solitary and those that are sociable. The difference is down to a single amino acid in a shared receptor. By changing the receptor it is possible to turn a solitary worm into a sociable worm.

12 | Alex, an African grey parrot, knew the meaning of words, and of shapes, colours and numbers. He understood concepts, including the concept of nothing. He would say 'Pay attention' if ignored. It took over twenty-five years of training on the part of his keeper, the animal psychologist Irene Pepperberg. He was still improving when he died, she said. Everything was just coming together in his mind. His last words to her were, 'You be good, see you tomorrow, I love you.' She is now training another grey parrot.

13 | Ayumu, a chimpanzee living in Kyoto, responds to patterns of numbers faster than humans can. A bonobo chimpanzee named Kanzi understands hundreds of words. Rico, a border collie, has been trained to remember hundreds of names for different objects (mostly soft toys, any of which he will retrieve on request).

A young vixen found injured, cared for and released back into the wild, turned up one day in the garden to show off her new cubs to Miriam. 'It was a breathtaking experience. I felt crowned.' But Miriam knew that the language of animals is the language of the soul, and this was a language she spoke as fluently as she spoke the cold language of science.
*Anthony Tucker's obituary of the naturalist and flea expert Miriam Rothschild (1908–2005)*

The greater part of Men make their way with the same instinctiveness, the same unwandering eye from their purposes, the same animal eagerness as the Hawk – the Hawk wants a Mate, so does the Man – look at them both they set about it and procure one in the same manner.
*John Keats*

Eventually [the cow] lost all respect for the boys and ran off. It was a great chase. They caught up with her in a trench half-way over the ridge, bridled her with cord, and led her home. She stood on the paving exhausted and forlorn, the veins in her neck swollen, her ears twitching in despair, nor did she stop complaining till Finna came out to her and stroked her and talked to her about life. When life is a weariness and escape is impossible, it is wonderful to have a friend who can bring us peace with the touch of a hand. After this Finna decided to tend to the cow herself. She took little Nonni with her. Those were good days. They were serene days and quite undemonstrative, like the best days in one's life; the boy never forgot them. Nothing happens; one simply lives and breathes and wishes for nothing, and nothing more.

*Halldór Laxness (1902–98), Independent People*

14 | One day in April 1915 at ten to eleven in the morning, the composer Jean Sibelius saw a flock of migrating Whooper swans – he counted sixteen of them – fly over his house. He said it was one of the greatest experiences of his life. The experience inspired the horn theme that appears in the finale of his 5th Symphony. When he had completed the score and laid down his pen, he went outside and saw swans circling his house, which he took to be a good sign.

The swans are always on my mind, and they lend magnificence to life. It is strange to note that nothing in the whole world, not in art, literature or music, has such an effect on me as those swans and cranes and bean geese, their calls and their appearance.

*Jean Sibelius (1865–1957), composer*

15 | When we see birds in flight we know for a moment that the world makes sense. I look out of my window and see the different species of birds flying and feeding together, and I feel lonely. Not lonely for other humans, but lonely for other species. Humans are social but isolated. We no longer live alongside a species close to our own. We may have lived alongside Neanderthals for millennia. I think of all our lost relations, all the hominoids, hominines, hominins, hominids that have come and gone. Do we remember somewhere deep in our bones what it was like to be gracile? Do we run marathons in order to try to re-create the sensation of running across the veldt in pursuit of an antelope?

My pilates teacher tells me to recall the gills I once had, and to imagine breathing through them. And then to breathe in as if through my skin – as if I were what, a bacterium? I think of when we were bacteria, not knowing then that the world has boundaries. Life was simpler then. Life was always simpler in the past.

All large-scale life was once worm-like, and still is: we are worms hung on skeletons, one long tube from mouth to anus, an exceptional long worm that folds in on itself into a long intestine that stretched out would reach to four times our height.

We can organise our common ancestors, put each inside the other, stretching back billions of years, and find to our surprise that everything that has ever lived fits into a worm, or even a bacterium. Further back, and everything that has ever existed fits inside a place of no dimension at the moment of the Big Bang.

Whatever we compare ourselves to we can choose to see the differences or we can see sameness: it depends at what scale, and with what lens we care to look. We see the human in apes when we look at the ape, but to call apes lesser humans is to demean both. We see apes in humans when we look at humans, but to say that humans are no more than apes is to demean both. With a different lens we see that human societies are like ant colonies or beehives, but they are also not at all like ant colonies or beehives. For as much as we might want to claim kinship with yeast, we are also clearly different from yeast. Might we even dare to say that we are more than yeast, or as strict Copernicans is that a step too far?

16 | Non-human primates are lonelier than we are, according to the primatologist Jane Goodall (b.1934). They are trapped within themselves. The philosopher Raymond Tallis says that even the most social animals lead essentially solitary lives compared to those of human beings. Humans have created a public arena of culture and art – in their widest meanings – in which to act out their lives. But do animals feel lonely? Our egos isolate us. Animals may not even know that they are separated from the outside world. Does a bird live in constant fear of being destroyed, or does the bird not know where it ends and nature begins? Prey sometimes enters a state of catatonia and gives itself up to

the predator. Or is it giving itself up to nature?* Perhaps loneliness is a quality of human life. Animals may be solitary, but solitariness is not loneliness.

Humans would be closed off from each other if it were not for language, emotion and culture. It is said that the private world of other animals is closed to us, but now that we are willing to accept that animals have emotions, make signs and bear gifts, surely this argument no longer has any validity. We used to think it was unscientific to talk of emotion in animals, but now that we understand that thinking begins in emotion, and only ends up in logic, scientists have become less cautious. We are more willing these days to acknowledge that it is grief we witness when crows mourn the death of a fellow crow. We don't know what the crows are thinking, but then neither do we ever know what other people are thinking, apart from what they tell us – unreliably – or what we gather from emotional and cultural cues.

Magpie funerals and chicken grief are steps too far for some. Research that suggests that insects can recognise human faces, or even count, downgrades what it means to recognise faces or to count, says Raymond Tallis. Animals behaving differently in the presence of different numbers of things has been taken as evidence of an ability to count, but it is humans that turn the ability seen in animals into counting. Animals do not have an innate grasp of abstract numbers. It is a hard-won ability of humans, and some primates. Chimps do not teach their young even how to break open a nut with a stone. The ability is innate. Similarly, sea-otters learn how to use stones placed on their stomachs to crack open clams, and hump-backed whales learn to escape nets – but they have to learn the skill individually, they do not know how to teach. Sometimes the circle of empathy may be stretched too far in our modish determination to accommodate other species. The animal behaviourist Marian Stamp Dawkins (b.1945) says 'that dogs and horses

---

* The explorer Dr Livingstone experienced this state when he was about to be attacked by a lion. (The lion was shot by his bearer.) There is presumably a chemical explanation, but it is hard to imagine what an evolutionary explanation would look like.

have famously fooled large numbers of people into thinking they could count when all they were doing was reading the body language of a human who was really doing the counting. Reading the body language of a counting person is a skill too, of course, but of a different kind; perhaps like the skill possessed by magicians, whom we take to be highly talented.

> As an atheist and also a humanist, I believe that we should develop an image of humanity that is richer and truer to our distinctive nature than that of an exceptionally gifted chimp.
> *Raymond Tallis*

> '... have a brandy-and-soda.' 'No more alcohol for me,' said Buffy. 'Look what it does to the common earthworm.' 'But you're not a common earthworm,' I said, putting my finger on the flaw in his argument right away.
> *P.G. Wodehouse (1881–1975), Summer Lightning*

17 | We might see ourselves in all animals, but it seems unlikely that any other species sees itself in us. Ants are not making documentaries, or watching them. We cannot land on the ceiling like flies, or dissolve into the background like squid, but we long ago escaped evolution driven by fitness to the environment. We make our own environments, and measure progress against the evolution of our technologies.

> Jacques Monod* used to say that everything that was true of *E. coli* was true of the elephant. But I don't think that even he said that everything that was true of the elephant was true of *E. coli*. I don't necessarily think the fly is as smart as Seymour [Benzer], even though Seymour doesn't know how to land on the ceiling.
> *Francis Crick, in a lecture at which the behavioural geneticist Seymour Benzer was present*

---

* (1910–76), biologist.

# SECTION 4

# On the relationship between human beings and other human beings

1 | The philosopher and biologist Herbert Spencer (1820–1903) invented the phrase 'survival of the fittest' to characterise Darwinian natural selection. Social Darwinism encouraged many Victorians to believe that the poor will always be with us. Social Darwinism led ultimately to the gas chamber.

> If a twentieth part of the cost and pains were spent in measures for the improvement of the human race as is spent on the improvement of breeds of horses and cattle, what a galaxy of genius might we not create!
> *Sir Francis Galton (1822–1911), polymath, cousin to Charles Darwin*

2 | Francis Galton's works started a vogue for eugenics in the United States. It was the mass-sterilisation programmes taking place in the US that first inspired the Nazis. The English biologist Julian Huxley (1887–1975) described eugenics as 'one of the supreme religious duties'.

3 | In response to a speech given by Julian Huxley at the Galton Dinner in 1936 at the Waldorf Hotel, Colonel Sir Charles Close said: 'Present-day Germany must be regarded as a vast laboratory which is the centre of a gigantic eugenics experiment ... It would be quite wrong and quite unscientific to decry everything which is now going on in that country

... The authorities there are in the position of being able to carry out the advice of their scientific advisers.'

4 | The American President Calvin Coolidge signed the Immigration Act in 1924: 'America must be kept American.' The intent was to bar immigrants from Africa and Asia. Nordic races were considered to be more consonant with American genes.

5 | The dancer Isadora Duncan in a letter to George Bernard Shaw: 'Will you be the father of my next child? A combination of my beauty and your brains would startle the world.' GBS replied: 'I must decline your offer with thanks, for the child might have my beauty and your brains.'

6 | As recently as 1970 Francis Crick advocated sterilisation for those 'poorly endowed' mentally.

I could not have believed how wide was the difference between savage and civilized man: it is greater than between a wild and domesticated animal.
*Charles Darwin, The Voyage of the Beagle*

The main conclusion arrived at in this work, namely that man is descended from some lowly organised form, will, I regret to think, be highly distasteful to many. But there can hardly be a doubt that we are descended from barbarians. The astonishment which I felt on first seeing a party of Fuegians on a wild and broken shore will never be forgotten by me, for the reflection at once rushed into my mind – such were our ancestors. These men were absolutely naked and bedaubed with paint, their long hair was tangled, their mouths frothed with excitement, and their expression was wild, startled, and distrustful. They possessed hardly any arts, and like wild animals lived on what they could catch; they had no government, and were merciless to every one not of their own small tribe. He who has seen a savage in his native land will not feel much shame, if forced to acknowledge that the blood of some more humble creature flows in his veins. For my own part I would as soon be descended from that heroic little monkey, who braved his dreaded enemy in order to save the life of his keeper, or from that old baboon, who descending from the mountains, carried away in triumph his young comrade from a crowd of astonished dogs – as from a savage who delights to torture his enemies, offers up bloody sacrifices, practises infanticide without remorse, treats his wives like slaves, knows no decency, and is haunted by the grossest superstitions.

*Charles Darwin, The Descent of Man*


7 | The differences Darwin observed between his own kind and those he called savages he took as evidence in support of natural selection. He believed that the savages would die out as being less fitted to the modern world, and his own kind thrive as a clear and separate species. The savages would eventually join all the other missing links that had died out and left the gaps that separate one species from another.


8 | Darwin observed that the sailors on the *Beagle* were better than the rest of those on board at picking out distant objects, but that Fuegians had even better eyesight than the crew. And yet when he described the Fuegians' reaction to shots fired from the *Beagle* he almost certainly misinterpreted their action: 'It was ludicrous to watch through a glass the Indians, as often as the shot struck the water, take up stones, and as a bold defiance, throw them towards the ship, though a mile and a half distant!' The archaeologist Timothy Taylor has suggested that by throwing stones they were not attacking the ship, but calculating its size. These were people who had never seen a ship before, and had no idea how large it might be. Was it a small object near the shore, or a large object in the distance? By throwing stones they could begin to work out the relationship between the size of the ship and its distance away.

9 | The earliest Tasmanians were clothed, used fire and ate fish, but by the time they were found by the outside world they had given up all three. Some commentators have seen these changes as deficiencies brought on by their isolation. But perhaps the changes made them 'more, well ... Tasmanian'.* It is only from our Western perspective that these changes look like backward steps. Fuegian aboriginals, as Timothy Taylor observes, were not illiterate, as many later European settlers were, but non-literate.

10 | In 1846 the Fijian chief Veidovi was captured and taken to New York; already ill, he died on the way. His head was sent to the Smithsonian Institution, where it can still be found, part of a collection of some 30,000 specimens of human remains.

11 | The last surviving full-blood aboriginal Tasmanian died in 1876. She was called Trugernanner. Syphilis had made her sterile. As she lay dying, she expressed her fear that her body would be exhibited as a zoological specimen. She asked to be buried near the spot where her mother had been murdered by whalers, or, in her last recorded words, 'behind the mountains'. Instead, she was buried at a former factory in a suburb of Hobart. Within two years her skeleton was exhumed, the bones boiled clean and wired together. She was placed on display at the Tasmanian Museum and Art Gallery near to the skeleton of a kangaroo. In 1976, close to the centenary of her death, her bones were cremated and the ashes scattered where she had wished them to be. Some of her hair and skin, which was in the collection of the Royal College of Surgeons in London, was returned to Tasmania for separate burial in 2002. The passing of this race was described by the anatomist Richard Berry (1867–1962), who made a study of Tasmanian aboriginal skulls, as a 'distinct step in human progress'.

---

* Timothy Taylor, *The Artificial Ape* (2010).

12 | In the 1930s the Nazis began systematically to destroy all evidence that Jews had ever existed. Did they suppose that there were edges and borders that might enclose the Jews as things, and who as things might be eradicated?

13 | History teaches us that every age is blind, and, being blind, does not know to what it is blind. Future ages will admonish us for what we do not see. Even Darwin, one of the most empathetic of humans, could not entirely escape the prejudices of his own age.* No one can. To understand the past is partly to unlearn what we know of the present.

14 | Victorian archaeologists who adopted Darwinism saw in the past a hierarchy of evolution that moved from savagery (hunters) to barbarity (farmers) to civilisation (city dwellers). Some might say that the distinction between hunters (savages), farmers (barbarians) and city dwellers (the civilised) survives today. To the Victorians a civilised society was only held back from barbarity by a class system, hence the Victorian obsession with the dynastic rule of Egyptian kings.

15 | As recently as 1906 the Australian anthropologist Sir Grafton Elliot Smith (1871–1937) wrote that 'the smallest infusion of Negro-blood immediately manifests itself in a dulling on initiative and a "drag" on the further development of the arts of civilization'.

---

* Darwin numbered among his valued advisers on variation his London barber Mr Willis – whom he quizzed about dog breeding while he had his hair cut – and his friend William Yarrell, a bookseller. He wrote their ideas down in his notebooks. He observed and recorded the development of his children. He had a gentle old horse called Tommy. He was a pigeon fancier. The years of travel on the *Beagle* meant far less to him than his years at home. At his home Down House he created a sand path that trailed about the gardens and that he planted with oak, lime and chestnut. He called it his thinking path. He walked it almost every day for forty years. He followed the bees around his garden from the violets to the primroses. It was here in his garden that he made his greatest leaps of imagination.

16 | In 1981 an attempt was made to claim the ancient Egyptians as part of an African story. The Senegalese historian Cheikh Anta Diop (1923–86) argued that 'Egyptians were Negroes, thick-lipped, kinky-haired and thin-legged'. Objectors pointed out that while some mummies show these features, the vast majority do not. The debate became acrimonious, some Afrocentrists insisting on a return to the name *Kemet* for Egypt.

17 | Sometimes we call what we do not understand primitive, sometimes magic. Each of us has a perspective to which we are attached. We only understand the other in relation to what we already know, but by believing, sympathising, even empathising, sometimes we can see beyond the limits of our knowledge. The alternative is to ignore or to attack. Societies have their own perspectives, so does the collectivism that is science. Solipsism may be impossible to escape, but at least the scientific method is an attempt to escape it.

SECTION 5

# On love

1 | The Greeks had a word for it. *Agape*, love of god, humanity; *Philia*, love of the pack, family; Platonic love, the chaste love that Socrates had for Alcibiades; *Eros*, lust and love of beauty; *Anteros*, requited love; *Pothos*, yearning and longing; *Himeros*, uncontrollable desire; *Kenotic* love, the highest version of the self, rejoicing eternally without the loss of personality.*

2 | For the ancient Greek philosopher Empedocles the universe is controlled by the tension between love and strife, Aphrodite and Ares. By the time Euclid wrote his propositions there was no longer room for love in cosmology. Because we do not know how to measure it?

3 | Is gravity, the inclination in all matter to draw closer together, physics' first pass at love? Is altruism biology's? Does love exist as a material force in the world? Can it be reduced to a set of responses? If not, then what is it?

I would lay down my life for two brothers, or eight cousins.
*The biologist J.B.S. Haldane (1892–1964), getting to the nub of the problem of mathematical descriptions of altruism*

---

* Their dissection of love could become minutely particular: *Katapepaiderastekenai*, to squander an estate through hopeless devotion to boys.

The political scientist Robert Axelrod explored the evolution of cooperation using game theory. He set up tournaments between computer programs playing the Prisoner's Dilemma, a game in which a player that betrays another gets a certain pay-off, and so on. The simplest strategy, which he called Tit for Tat but you might call Do unto others as they have just done unto you: a smile for a smile, a slap for a slap; wins. I don't think that's an accident. Fundamentally, in the end, cooperation wins out over selfishness … There's a sort of law of the universe that says cooperation is a very good strategy. It's sort of a law about the way the world works.

*Douglas Hofstadter, cognitive scientist*

4 | In the early 1960s, the biologist Bill Hamilton formulated an equation that describes kin selection. E.O. Wilson and others claimed that Hamilton's equation had, by reducing it to the mathematics of genetic relatedness, solved the problem of altruism. Decades later, Wilson became critical of his earlier stance. He now says that the mathematics of kinship rarely works.* It works for ants, but not for higher social animals like human beings. There are human qualities that the Hamilton equation cannot explain. Humans make sacrifices in complex ways, and for reasons that fall outside the power of this equation. Of any equation? Humans fall outside both the robotic determinism of ants and the selfish determinism of the individual.

5 | Neuroscientists V.S. Ramachandran and Antonio Damasio have heralded the discovery of mirror neurons as one of the most important neurological advances of modern times. Their discovery brings us closer to a physical explanation of empathy. Mirror neurons are the neurons that fire when we see someone else doing something, and then fire again when we do the same thing ourselves. Critics point out that no causal connection has yet been established. Mirror neurons were

---

* Wilson was heavily criticised for his early support of Hamilton. Now that Hamilton's equation is more favourably regarded, Wilson finds himself heavily criticised again.

discovered in monkeys in the late twentieth century. Ramachandran has subsequently found comparable neurons in the premotor cortex of humans. Mirror neurons might finally explain the contagion of yawning, a trivial but persistent phenomenon lacking an explanation; that they might explain empathy is highly contested.

> Parasitism may be a step toward brotherhood and even love, often becoming about as good in the long run for the host as for the parasite, both of whom tend to become dependent on each other under their unwritten but well-understood contract, as surely as you rely on your intestinal bacteria for digestion and vitamins or on your cat to suppress your mice.
> Guy Murchie, *The Seven Mysteries of Life*

> Blindness to suffering is an inherent consequence of natural selection. Nature is neither kind nor cruel but indifferent.
> *Richard Dawkins*

6 | To a hard-line reductionist love is an inducement that encourages us to propagate our genes.

7 | At what point does an act become so altruistic that it becomes indistinguishable from love? Is it when you would do something for the pack no matter what the cost? But surely we love heroes because they are not calculating. The hero strives to save a perfect stranger. The hero risks the humiliation of failure. The hero, for the sake of another and by risking death, stumbles on his highest self. Heroism cuts across prejudice and biological weightedness. Acting as if the miraculous exists, to risk death even though logic tells you that you will die, perhaps this is heroism.

Fancying, fertility and ... sexual behaviour – are not the whole story. Conversation, common interests, shared tastes, deep sympathy ... play a part.
*Raymond Tallis*

8 | Sometimes to be in love is about not being in love at that particular moment. How can we hope to find love in an fMRI scan?

Formulas are nothing. Life is everything. And life is simultaneously mind and heart.
*Eileen Gray (1878–1976), furniture designer and architect*

You mean more to me than any scientific proof.
*A line from Andrei Tarkovsky's film Solaris (1972)*

The fact that buildings don't fall down and you can eat unpoisoned food that someone grew, is immediately palpable evidence of an ocean of goodwill and good behaviour from almost everyone, living or dead. We are bathed in what can be called love.
*Jaron Lanier*

If I really love the world I will fit myself to it, rather than make it fit me.
*Thomas Mann, Confessions of Felix Krull*

What love is really about is a bottomless empathy, born out of the heart's revelation that another person is every bit as real as you are.
*The novelist Jonathan Franzen, in an interview*

The Emperor Conrad III having besieged Guelph, Duke of Bavaria,* would not be prevailed upon, what mean and unmanly satisfactions soever were tendered to him, to condescend to milder

---

* In 1140, in Weinsberg, Upper Bavaria.

conditions than that the ladies and gentlewomen only who were in the town with the duke might go out without violation of their honour, on foot, and with so much only as they could carry about them. Whereupon they, out of magnanimity of heart, presently contrived to carry out, upon their shoulders, their husbands and children, and the duke himself; a sight at which the emperor was so pleased, that, ravished with the generosity of the action, he wept for joy, and immediately extinguishing in his heart the mortal and capital hatred he had conceived against this duke, he from that time forward treated him and his with all humanity.

*Montaigne*

I am of a constitution so generall, that it consorts, and sympathizeth with all things; I have no antipathy, or rather Idio-syncrasie, in dyet, humour, ayre, any thing; I wonder not at the French, for their dishes of frogges, snailes, and toadstooles, nor at the Jewes for Locusts and Grasse-hoppers, but being amongst them, make them my common viands; and I finde they agree with my stomach as well as theirs; I could digest a Sallad gathered in a Church-yard, as well as in a Garden. I cannot start at the presence of a Serpent, Scorpion, Lizard, or Salamander; at the sight of a Toad, or Viper, I finde in me no desire to take up a stone to destroy them. I feele not in my selfe those common antipathies that I can discover in others: Those nationall repugnances doe not touch me, nor doe I behold with prejudice the French, Italian, Spaniard, or Dutch; but where I finde their actions in ballance with my Countrey-mens, I honour, love, and embrace them in the same degree; I was borne in the eighth Climate, but seeme for to bee framed, and constellated unto all; I am no Plant that will not prosper out of a Garden. All places, all ayres make unto me one Country; I am in England, every where, and under any meridian; I have beene shipwrackt, yet am not enemy with the sea or winds; I can study, play, or sleepe in a tempest.

*Thomas Browne (1605–82), Religio Medici*

I like so many things, fairy stories, fat butlers, porters, the smell of tangerines, suave Orientals, a good tune, lovely colours, French accents, puppies, bath salts ...
*Elizabeth Bowes-Lyon (the future Queen and Queen Mother, here when Duchess of York), in a letter*

A Poet is the most unpoetical of anything in existence; because he has no Identity – he is continually ... filling some other Body ... It is a wretched thing to confess; but is a very fact that not one word I ever utter can be taken for granted as an opinion growing out of my identical nature – how can it be, when I have no nature? When I am in a room with People if I ever am free from speculating on creations of my own brain, then not myself goes home to myself; but the identity of every one in the room begins so to press upon me that I am in a very little time annihilated.
*John Keats, in a letter dated 27 October 1818*

I am certain of nothing but of the holiness of the Heart's affections and the truth of the Imagination.
*John Keats, in a letter dated 22 November 1817*

If we could read the secret history of our enemies, we should find in each life sorrow and suffering enough to disarm any hostility.
*Henry Wadsworth Longfellow (1807–82), poet*

... meaning by friendship the frank unreserve, as before another human being, of thoughts and sensations; all that objectless and necessary sincerity of one's innermost life trying to react upon the profound sympathies of another existence.
*Joseph Conrad, Nostromo*

Love lets others be, but with affection and concern.
*R.D. Laing (1927–89), psychiatrist*

9 | Plato imagined that humans had once existed as more symmetrical and spherical creatures that became divided into two separate beings. We still talk of our 'other half', as if he or she was lost and is now found.

The sun was setting in front of us in a blaze of pink and golden light. His Highness waved a regretful hand towards it. 'I want a friend like that,' he said.
*J.R. Ackerley (1896–1967), Hindoo Holiday*

How should we like it were stars to burn
With a passion for us we could not return?
*W.H. Auden (1907–73), poet*

Doubt thou the stars are fire;
Doubt that the sun doth move;
Doubt truth to be a liar;
But never doubt I love.
*Shakespeare, Hamlet*

10 | 'Nothing can penetrate the loneliness of the human heart except the highest intensity of the sort of love the religious teachers have preached,' are the words not of some mystic, but of that famous atheist Bertrand Russell, here recognising the power of faith even as he himself is unable to profess it. Russell had seen too many Christians sucking life dry for him ever to want a part in it, but his words are full of spiritual regret and longing.

11 | The unconsummated love of Petrarch and Laura. The unconsummated love of Michelangelo and a much younger man. The troubadour's unconsummated love for the distant object. The saint's love – eternally unconsummated – for that most distant object of desire, God.

12 | Love God, said St Augustine, and you may do as you please.

As after a single drop is poured from a bottle its contents pour out in great streams, so in my soul my love for Varenka released all the capacity for love concealed in it. At that moment I embraced the whole world with my love.

*Leo Tolstoy, 'After the Ball'*

He devoted a lot of thought to the intricate workings of the heart. He observed consciously and unconsciously the effects of beauty on the imagination, how impressions grew into feelings and the effect they had in their turn and how they played themselves out. As he entered life he looked around him and arrived at the conviction that love, like the lever of Archimedes, was powerful enough to move the world and that love, properly understood and used, was as much a force for universal, absolute truth and good as, perverted and misused, it was a force for falsehood and evil. But where was good and where was evil and where was the line between them?

*Ivan Goncharov, Oblomov*

Love can do all but raise the dead.

*Emily Dickinson (1830–86), poet*

Love is space and time measured by the heart.

*Marcel Proust*

Love can damn you for all eternity. Love will take you to hell. But if you are lucky, if you are very lucky, it will bring you back again.

*Mark Rowlands, The Philosopher and the Wolf*

And we got drunk on alcohol
And on love the strongest poison and medicine of all.

*Joni Mitchell, singer-songwriter*

13 | Under stress – sadness, anxiety, experience of death – immune cells become coated with chemicals and disabled. We can die of a broken heart.

Being in love (at least this was how she imagined it) was something for people who were determined to get the better of each other, a cruel, hard sport – far crueller and harder than tennis! – with no holds barred, with all kinds of low blows, and without any concern, just to palliate things, for the good of the soul or for notions of fairness.
*Giorgio Bassani (1916–2000), The Garden of the Finzi-Continis*

14 | Nietzsche saw love as divided: Apollo and Dionysus, domestic and sexual loves that are hard if not impossible to combine in the one object of desire.

15 | Love is indifferent. It does not care whether it creates or destroys. Nature is indifferent but we are not; and being part of nature, we are the part of nature that is not indifferent. Nature removes the weak, but we do not have to. Our relationship to the weak and powerless is what makes us human at our best and worst. We are, after all, but for our minds, the weak species that should not have survived.

Love looks not with the eyes but with the mind.
*Words tattooed on the fashion designer Alexander McQueen's upper right arm, taken from Shakespeare's A Midsummer Night's Dream*

16 | The strongest electrical signal to the brain comes from the heart.

I believe in love; not just getting it, but giving it, no matter whether you're loved in return. It doesn't matter so long as you love someone.
*Lines from Robert Altman's film Gosford Park (2001), spoken by Dorothy (played by Sophie Thompson)*

If equal affection cannot be,
Let the more loving one be me.
*W.H. Auden*

'Life is, alas, so badly arranged that we rarely enjoy that happiness. Mme de Sévigné was actually better off than most: she spent much of her life with a loved one.'

'You're forgetting that it wasn't love, though – it was just her daughter.'

'But the important thing in life is not whom one loves,' he exclaimed, in a voice that was authoritative, peremptory, almost cutting. 'The important thing is to love. The feelings of Mme de S. for her daughter can more properly deserve the name of the passion depicted by Racine in *Andromaque*, or *Phèdre* than the paltry goings on between the young Sévigné and his mistresses. The same goes for the love of a mystic for his God. The limits we set to love are too restrictive and derive solely from our great ignorance of life.'
*Marcel Proust, Remembrance of Things Past*

It is possible to carve out a life based on love, which after all, is the only life worth living.
*Derrick Jensen, naturalist*

Wisdom, knowledge in the broadest sense, is always motivated by love.
*David Orr, environmentalist*

# SECTION 6

# On doing the right thing

1 | 'Never impose on others what you would not choose for yourself.'
'What is hateful to yourself, do not do to your fellow man.' 'All things
whatsoever ye would that men should do to you, do ye even so to them.'
The Golden Rule can be found in the New Testament, the Talmud, the
Koran, Confucius and Socrates.

> To do good in return for evil, to love your enemy, is a height of
> morality to which it may be doubted whether the social instincts
> would, by themselves, have ever led us. It is necessary that these
> instincts, together with sympathy, should have been highly culti-
> vated and extended by the aid of reason, instruction, and the love
> or fear of God, before any such golden rule would ever be thought
> of and obeyed.
> *Charles Darwin, On the Origin of Species*

2 | The liberal view of morality hasn't changed much since Socrates
(c.469–399 BC): that human reason provides sufficient means for
working out what is moral and what are good actions. Socrates said that
knowledge of good and evil is innate. Plato (428–348 BC) added that
we need to be trained in order to realise our capacity to understand
what is moral and what are good actions. The Roman Emperor and
Stoic philosopher Marcus Aurelius (AD 121–80) argued that we should
live good lives as a means of advancing the common good of society. St
Augustine (354–430) said revelation was needed in addition to reason,

because we are imperfect and need the assistance of God. Humanism began to flower in the fourteenth century with the poet Petrarch. It is the belief that humans are good and perfectible, and it embodies an idea of progress. Kant said that what we want to do is in conflict with what we know we ought to do; and because they are not the same, what we know we ought to do can be taken to be the will that forces reason to accept God on faith. Moral law, he said, is our private imitation of God's enforcement of physical law. David Hume thought our will was instead 'our feeling for the happiness of mankind', a kind of altruism. The political economist John Stuart Mill (1806–73) said that will alone is not enough, and needs feeding by religion, education, institutions, society and culture. Mill, like Marcus Aurelius, was concerned with living a good life in society. We have a desire for unity with others, but we also need to develop our conscience – which is a learned attribute. For Darwin the human mind is different only in degree, not in kind, from the workings of the brains of other animals. Modern Darwinists emphasise inherited characteristics, but Darwin himself was open to cultural agencies of change.

Modern philosophers have argued that what matters is not absolute morality but what is best for the group. Our idea of goodness arises out of what happens to societies that adopt different strategies. What happens to, say, a totalitarian regime is a test of our ideas of goodness. If we believe in liberalism we are relieved when totalitarian regimes – as a test of our beliefs – ultimately fail.

Perhaps Jeremy Bentham's (1748–1832) philosophy of utilitarianism, doing the greatest good for the greatest number, is our best shot at a moral code for our times. The secret channels through which goodness flows about the world being secret, we do not see, nor know truly, if they exist; but by believing in them we might achieve what we hope they effect whether they exist or not. We cannot know that good causes produce good effects, but if we are to believe anything, surely we might take up this as the most modest of faiths. Yet even utilitarianism has its dark side. When pure logic argues that individuals – perhaps the disabled or the old – must be sacrificed for the common good, even this seemingly beneficent philosophy can turn into a kind of extremism.

3 | Human beings transcend mere logic. They are inconsistent and hold opposites together. There are logical answers, and there are human ones.

I think we realise too little how often our arguments are of the form: A says 'I went to Grantchester this afternoon' and B argues in reply, 'No, I didn't.'
*Frank Ramsey (1903–30), mathematician*

'Contrariwise,' continued Tweedledee, 'if it was so, it might be; and if it were so, it would be: but as it isn't, it ain't. That's logic.'
*Lewis Carroll, Through the Looking-Glass*

'*Marry* ... Children ... Constant companion ... better than a dog anyhow ...

*Not Marry*: My God, it is intolerable to think of spending ones whole life, like a neuter bee, working, working, & nothing after all. No, no won't do ...'

And?

'Marry – Marry – Marry QED ... It being proved necessary to Marry. When? Soon or Late ... already beginning to wrinkle ...'
*On a scrap of paper, Darwin worrying over the pros and cons of marrying Emma Wedgwood*

I had learned something in marrying Denise. I had been reluctant and fearful of marriage, even to Denise, whom I loved much more than any other woman I had ever thought of marrying. But Denise was confident that our marriage would work, so I took a leap of faith and went ahead. I learned from that experience that there are many situations in which one cannot decide on the basis of cold facts alone – because facts are often insufficient.
*Eric Kandel, biologist*

4 | A thirty-year-old man and a baby are trapped in a burning car. There is time only to rescue one of them. Which? The answer is logically clear. The baby has little to lose. According to the moral psychologist Joshua Greene (not entirely seriously, I think) the man has 'stronger and further-reaching desires for more life than does the baby'.

Greene devised an experiment in which the subjects are given the choice: either physically push a person in front of a train to save five people, or throw a switch that would activate a machine that would push that same person. Almost everyone thought it was wrong to push but OK to throw the switch.

Human beings reduced to laboratory conditions seem to behave as if they belong to some fascistic state, perhaps because fascism – from the perspective of those of us who live in a democracy – feels unreal and artificial.

5 | Exceptions have tremendous force in science but less force in collective human existence. Only in extremist regimes are we required to make such choices as are devised by behaviourists. The world is contingent. Our lives are shaped by what has come before. We live in the shared past of everyone who has ever lived. In dictatorships history is partitioned into a false present cut off from the context of the past. Dictatorship is pure logic as government.

6 | Keynes wrote that World War II was an attempt to bring Germany 'back into the historic fold of Western civilisation of which the institutional foundations are the Christian Ethic, the Scientific Spirit and the Rule of Law. It is only on these foundations that the personal life can be led.'

> ... since good, the more
> Communicated, more abundant growes.
> *John Milton*

The something not ourselves that makes for righteousness.
*Matthew Arnold (1822–88), poet*

We are the playthings of the gods and so should play the noblest games.
*Plato*

Do not despair, one of the thieves was saved; do not presume, one of the thieves was damned.
*St Augustine*

For the growing good of the world is partly dependent on unhistoric acts; and that things are not so ill with you and me as they might have been, is half owing to the number who lived faithfully a hidden life, and rest in unvisited tombs.
*The final words of George Eliot's Middlemarch*

We are all responsible for what the future holds in store. Thus it is our duty, not to prophesy evil but, rather, to fight for a better world.
*Karl Popper (1902–94), philosopher*

War, like any other dramatic spectacle, might possibly cease for want of a 'public'.
*George Eliot, The Mill on the Floss*

Our business is to continue to fail in good spirits.
*Robert Louis Stevenson (1850–94), novelist*

'Vive la bagatelle!'
The Man that loves and laughs, must sure do well.
*Jonathan Swift*

You have to be somebody before you can share yourself.
*Jaron Lanier*

It is a fine thing to be independent in life, and a proud sensation to know yourself unique: but a person who stands all on his own, utterly detached from his fellows, may come to feel that reality itself is an illusion.
*Jan Morris, Conundrum*

# On the difficulty of being

1 │ It is tempting to suppose that no other species knows so much about the world as we do, but perhaps it is more accurate to say that no other species has worked out how to systematise its knowledge of the world, and that is why we rule it. But we know very little about how to be in the world, very little about what our motives are. We have very little self-knowledge. We know ourselves least of all. We are full of inconsistencies and incongruities, hardly able to control, understand or even acknowledge our needs and drives. It is possible that other animals – perhaps most other animals – are better at being in the world than we are.

> Freud's word for the instinct-driven unconscious part of his tri-unal self was the It, das Es, which became 'id' in English, a term he borrowed from Georg Groddeck, who wrote, 'I hold the view that man is animated by the Unknown, that there is within him an "Es", an "It", some wondrous force, which directs both what he himself does and what happens to him. The affirmation, "I live" is only conditionally correct, it expresses only a small and superficial part of the fundamental principle, "Man is lived by It."'
> *Siri Hustvedt, The Shaking Woman*

When I look out on a night such as this I feel as if there could be neither wickedness nor sorrow in the world; and there would be less of both if the sublimity of Nature were more attended to, and people were more carried out of themselves by contemplating such a scene.
*Fanny Price, in Jane Austen's Mansfield Park*

Scenery is fine – but human nature is finer.
*John Keats, in a letter*

Human life – that appeared to him the one thing worth investigating.
*Oscar Wilde, The Picture of Dorian Gray*

The great effort of construction that is living.
*Clarice Lispector, The Passion According to G.H.*

A vale of soul-making.
*John Keats*

Human relationships aren't nearly as complicated as people make out: they're often insoluble but only rarely complicated.
*Michel Houellebecq, Platform*

We are born mad, develop a conscience and become unhappy; then we die.
*David Eder (1865–1936), psychologist*

2 | Madness, or at least the possibility of it, creeps up on us when we are alone too long. It is as if other people contain us and keep us sane; and that without them we are liable to flow out of ourselves, messily.

If we were not all so interested in ourselves, life would be so uninteresting that none of us would be able to endure it.
*Arthur Schopenhauer (1788–1860), philosopher*

Being alive is like getting a gift you didn't ask for. You can return it, but that's in poor taste and hurtful.
*Wendy Wasserstein (1950–2006), playwright*

'Anyway – doesn't matter.' One may hear this statement, which is analogous to a reflex, spoken by all who have a touch of self-esteem, in circumstances which can vary from the trivial to the tragic, and which reveals, as it did on the present occasion, how much the thing which is said not to matter does matter to the speaker; and in the tragic vein, the first thing to come to the lips of any man who takes a certain pride in himself, if his last hope has just been dashed by someone's refusal to help him out, may well be the brave, forlorn words: 'Oh well, it doesn't matter, not to worry – I'll think of something else,' the something else which is the alternative to what 'does not matter' being sometimes the last resort of suicide.
*Marcel Proust, Remembrance of Things Past*

We are a race with conscience enough to feel that it is vile, and intelligent enough to know that it is insignificant. We survey the past, and see that its history is of blood and tears, of helpless blundering, of wild revolt, of stupid acquiescence, of empty aspirations. The energies of our system will decay, the glory of the sun will be dimmed, and the Earth, tideless and inert, will no longer tolerate the race which has for a moment disturbed its solitude. Man will go down into the pit, and all his thoughts will perish.
*Arthur Balfour,\* The Foundations of Belief*

---

\* In a Prime Minister we might desire a less cosmic perspective. 'Nothing matters very much,' Balfour (1848–1930) once said, 'and few things matter at all.' With such a

At the Times Square subway station in New York is a seemingly endless tunnel that joins 7th and 8th Avenues. At intervals along the tunnel, in capital letters just above head height, you may read the Commuter's Lament by Norman B. Colp:

Overslept,

So tired.

If late,

Get fired.

Why bother?

Why the pain?

Just go home

Do it again

We, like everything, are driven by purposeless decay.
*Peter Atkins, chemist*

The firm foundation of unyielding despair.
*Bertrand Russell (1872–1970), philosopher*

3 | Freud said of psychoanalysis that he hoped to transform hysterical misery into common-or-garden unhappiness.

Only unhappiness is elevating, and only the tedium that comes from unhappiness is heraldic like the descendants of ancient heroes.
*Fernando Pessoa, The Book of Disquiet*

That daily dose of poison, recently invented, that we call happiness.
*Coco Chanel (1883–1971), fashion designer*

---

philosophy, it is not surprising that in assessing his legacy the journalist Harold Begbie wrote that, 'To look back upon his record is to see a desert, and a desert with no altar and with no monument, without even one tomb at which a friend might weep.'

4 | When we first meet Madame Merle* she is seen from behind, sitting at a piano playing a piece that Isabel Archer cannot quite place. Isabel waits for the music to finish and for this unknown and unexpected guest to reveal herself. In just a few sentences, out of the blankness and possibility of a back, and a piano expertly played by an interloper, James somehow manages to establish Madame Merle's mystery and her sinister controlling passivity. Madame Merle – even her name is disquieting – turns to Isabel, and her first words amount almost to a philosophy: 'I'm afraid there are moments in life when even Schubert has nothing to say to us. We must admit, however, that they are our worst.' It would be a refined existence indeed in which Schubert arbitrates our mood rather than, say, the need to earn money, or the zealot at the door with a machete. I had a friend who fainted in the face of too much beauty. He had been brought up by aunts, his sensibilities tuned to fever pitch from an early age. A world of Jamesian refinement may not be for everyone. But there are many worlds, and we might agree that too much refinement is better than none.

5 | On an idyllic summer's day the then provost of King's College, the philosopher Bernard Williams, invited a group of friends on a picnic. Kitty Godley (daughter of the sculptor Jacob Epstein), whose favourite painters were Crivelli and Fantin-Latour, excused herself in the middle of lunch, saying: 'This is too beautiful. I just have to leave.'

> She wanted to lie in a hammock beside a blue tideless sea and think about Tibullus.
> Ford Madox Ford, *Parade's End*

6 | There are days when the world is a veil and days when it is a rock. There are yawning days when yet again Rochester does not return, and

---

* In Henry James's *The Portrait of a Lady*.

there is nothing to do but stare out of the window at the rain. There are days when like N for Neville* we might die of ennui.

> I know no one but you who can be fully sensible of the turmoil and anxiety, the sacrifice of all what is called comfort the readiness to Measure time by what is done and to die in 6 hours could plans be brought to conclusions – the looking upon the Sun the Moon the Stars, the Earth and its contents as materials to form greater things – that is to say ethereal things – but here I am talking like a Madman greater things that our Creator himself made!!
> *John Keats, in a letter to Benjamin Robert Haydon, 10–11 May 1817*

7 | Without scientific progress there would be no refined life indoors from which, should we care to, to disparage progress. Comfort grants us the luxury to suffer existentially, which is preferable to suffering physically. Only on our most pampered and pessimistic days would we claim that humans have not progressed. Life is better when we are protected from the weather and unwelcome parasites. We may not be free, nor know how to deal with our freedom, but we can hardly blame scientific progress for that.

> It's rather a strong check to one's self-complacency to find that much of one's right doing depends on not being in want of money.
> *George Eliot, Middlemarch*

8 | From a Darwinian perspective, it is clear that it is harder to do good when the struggle for survival dominates life. Logically, it could be argued that to do good when living conditions are harsh (or to do nothing towards the general good when those conditions are fair) might constitute criteria for assessing moral worth. It was why Henry James chose to write about the rich. The poor might be excused bad behav-

---

* From *The Gashlycrumb Tinies* by Edward Gorey.

iour, whereas observing the rich behaving well or ill was a way of scrupulously investigating scientifically purer specimens.

Intellectual freedom depends upon material things.
Virginia Woolf, A Room of One's Own

9 | It is hard to believe in anything in our cynical age. I know that I live, as most of us do in the West, a privileged life. I know I am protected from the wildness of nature because of technological affirmations of the scientific method. I know I am also protected by relative wealth, by democratic liberalism and peace (or at least the illusion of them). I know that the main obstacles to freedom are material: the shortage of fresh water, of enough to eat, of proper housing. I know that in the world elsewhere poverty, disease, war and tyranny bring unremitting suffering to millions. I know that not to be grateful for the life I have is to dishonour the lives of the millions who died in wars where personal freedom found itself challenged by dictatorships. And if, for a new generation, the weight of that past begins to fade along with the collective memory of the world wars, a new oppressive atmosphere rushes in to fill the vacuum. The life we fear we do not sufficiently appreciate, but to which we are nevertheless addicted, has come at the cost of a fading future. We find ourselves in an apocalyptic age, in which we are in danger of denying the wonder and difficulty and uniqueness of being human beings. And yet is our age any different from 'The dark dawning of our modern day,' that Livy wrote of in his History of Rome, 'when we can neither endure our vices, nor face the remedies needed to cure them'? Life as we find it, Freud once said, is too hard. And it always has been. The first noble truth, said the Buddha, is that life is suffering. These are not necessarily apophthegms of despair; they can be seen as declarations of hope. The Buddha contemplated the nature of suffering in order to understand how it might be transcended. Christ suffered so that we might not. Kindness is the highest human quality, says the Dalai Lama; charity, said Christ.

What is the meaning of life? That was all – a simple question; one that tended to close in on one with years. The great revelation had never come. The great revelation perhaps never did come. Instead there were little daily miracles, illuminations, matches struck unexpectedly in the dark ...
*Virginia Woolf, To the Lighthouse*

A human self ... has a strong tendency to maintain a constant form and it accepts only those new experiences that are appropriate to its form and rejects those that are not. Normal human behaviour ranges between the extremes of nervous breakdown, on the one hand, which occurs when incoming experiences disrupt the self due to their intensity or the fragility of the self, and catatonic states, at the other extreme, which occur when the self maintains its integrity by refusing to respond to new inputs. Most humans live between these two extremes, work hard to maintain the structural integrity of the self and modify it only under great pressure and with great reluctance ... Such rare changes in selves are called 'conversions' or crises experiences.
*Manuel Davenport, The Mystery of Morality*

10 | The psychologist Dorothy Rowe (b.1930) wrote that anxiety and depression are what follow when our world view is challenged and we find that we no longer fit the universe as we find it to be.

I know of nothing more difficult than knowing who you are and having the courage to share the reasons for the catastrophe of your character with the world.
*William Gass, novelist*

And this, of course, was the simplest definition of depression that he knew of: strongly disliking yourself.
*Jonathan Franzen, Freedom*

The mere thought of having to enter into contact with someone else makes me nervous. A simple invitation to have dinner with a friend produces an anguish in me that's hard to define.
*Fernando Pessoa, The Book of Disquiet*

I don't know if I'm unhappy because I'm not free, or if I'm not free because I'm unhappy.
*Patricia Franchini in Jean-Luc Godard's film À bout de souffle*

This melancholy of yours, this brooding ... is really a sign of strength. Sometimes an active questioning mind tries to probe beyond normal limits and, of course, finds no answers, and that's when the melancholy sets in ... a temporary dissatisfaction with life ... a deep-seated frustration with life for not yielding up its secrets.
*Ivan Goncharov, Oblomov*

What can the word 'depressed' mean to the depressed? It is no more than the echo of the patient's voice.
*John Berger, A Fortunate Man*

11 | About a decade ago I began to suffer panic attacks. They were frequent but not regular. I could not predict when the next one would arrive nor what might precipitate it. At that time in my life I only admitted consciously as guides to decision-making whatever I could work out logically or rationally. Clearly my first concern was to have my condition diagnosed. The first doctor told me that I was depressed and put me on a combination of Xanax (an anti-anxiety drug) and Cipramil (a serotonin enhancer). The latter drug initially had the effect of making my panic attacks so acute that for a few days there was the possibility that I might have to check into a clinic in order to be eased onto the drug (unlike most of the patients at this somewhat famous establishment who were there to be weaned off the drugs to which they had become addicted).

There must have been curious firing patterns in my brain. I felt that whoever I used to be had fled. Someone was left behind, but not anyone I recognised. I began to understand what the horror of madness might be like. When there is nowhere to go to, when all the exit routes are blocked, life is a prison in which one has become not so much imprisoned as the prison itself. Whatever had precipitated what I had become, I was certain that chemical treatment for depression would not get to the heart of the matter.

The next doctor told me that I was pre-diabetic and that I would need to change my diet for the rest of my life. I had become highly sensitive to anything sugary or that soon changed to sugars. I could begin to predict that twenty minutes after eating pasta a panic attack was in the offing. But then other things also seemed to trigger an attack; not just food, thoughts too. I understood what I had not understood, that there is no boundary between mind and body, that a thought is as powerful a physical agency in the world as any material object. It was as if my body had been retuned, becoming now as reactive to a thought or the ingestion of a biscuit as it might once have been to a blow to the head.

When it is your own life or sanity that seems to be at stake, collectivism, I discovered, goes out of the window. Traditional medicine was not working, and I had to look elsewhere, the elsewheres being places I would have scorned from the smug heights of my well-being.

12 | As soon as I was returned to wellness, as soon as the self I thought had fled forever returned home, I began to justify my experiences in the old ways. There are no supernatural powers. In medicine, whatever happens supranaturally can be explained by the placebo effect.

'The doctor will see you now' was once a spell. To be looked at intently is rare and powerful. To be a good physician maybe requires some of the qualities required of a good magician, to notice tics and read signs. These days most doctors do not have the time even to look up from their computer screens. Perhaps what I was paying for when I looked for help outside the boundaries of orthodox medicine was the belief radiated from another human being that I would be well. Which is something.

As Nicholas Christakis, a medical sociologist, has pointed out, many commonly used remedies, such as Viagra, work less than half the time, and there are conditions, such as cardiovascular disease, that respond to placebos for which we would never contemplate not using medication, even though it proves only marginally more effective in trials. Some patients with Parkinson's respond to sham surgery.
Louis Menand, New Yorker *magazine*

Many alternative practitioners develop an excellent relationship with their patients, and this helps to maximise the placebo effect of an otherwise useless treatment.
Edzard Ernst, *professor of complementary medicine, in the New Scientist*

13 | Some placebo studies appear to undermine the foundations of evidence-based medicine. One study showed that nearly 80 per cent of the effect of antidepressants derives from faith that they will work. The effect of belief may be even stronger in alternative medicine. Another study – conducted among patients with Parkinson's disease – showed that a certain drug increased the secretion of the neurotransmitter that the disease depletes. This could be taken to be evidence in support of the efficacy of the drug, except that even those in the control group produced more of the same neurotransmitter, and at the same levels; presumably because they believed that they were not in the control group.

14 | It has been suggested that patients have now become so know-ledgeable about the nature of control groups that they can tell if they are in a control group or not, simply by noting that if they do not experience any side effects they must be taking a placebo; knowledge which

is enough to undermine the validity of the control. But the complications do not rest there: some studies have found that even knowing you are receiving a placebo doesn't necessarily reduce its effectiveness.

> It is in moments of illness that we are compelled to recognize that we live not alone but chained to a creature of a different kingdom, whole worlds apart, who has no knowledge of us and by whom it is impossible to make ourselves understood: our body.
> *Marcel Proust*

15 | Descartes desired 'the general well-being of men' (a desire that came, unfortunately, at the expense of the well-being of all other animals). He believed that advances in medicine would have the greatest influence on human happiness, and make men 'wiser and more dexterous'. He tried to turn medicine into a deductive science, but found it harder even than turning ethics into a deductive science.

# On dreams and doing nothing

Dreams are real while they last. Can we say more of life?
*Havelock Ellis (1859–1939), social reformer*

The royal road to the unconscious.
*Sigmund Freud, of dreams*

The courage to go toward the greatness of sleep.
*Clarice Lispector, The Passion According to G.H.*

Only my ghostly and imaginary friends, only the conversations I have in my dreams, are genuinely real and substantial.
*Fernando Pessoa, The Book of Disquiet*

Besides, it's not as though I was sleeping – I've been reading, scribbling some notes for my thesis, and looking out of the window. I always do a whole lot of things when I'm in bed. The warmth of the blankets undoubtedly spurs me into activity.
*Giorgio Bassani, The Garden of the Finzi-Continis*

I believe in the dream. I think we only live for our dreams and in our imaginations. That's the only reality we ever know.
*Diana Vreeland (1903–89), fashion editor*

1 | Perhaps the point of being awake is to realise our dreams; not just the daydreams of the ego but the sublime dreams that haunt our sleep.

2 | There are researchers who claim that dreams serve no purpose and that sleep is merely a pit stop, a time to drop, for as short a period as possible, out of consciousness in order to service the unconscious body for the real work of the daylight hours. For them, dream chasers must seem like indulgent time-wasters. I'll sleep when I'm dead, some people say, as if sleep and extinction were two equal absences. The dead and the sleeping may share a passing resemblance, but sleep is not annihilation. What I remember from before my birth is nothing at all, not some dreaming pre-existence. To degrade sleep is to write off a large part of life, as if being conscious is the only point to living. Proust loved to sleep, of course he did. Through the portal of sleep – dreaming and dreamless – he dropped down into the wisdom of his body. Even when we are awake, most of what we are is unconscious. A life of perpetual torpor and indolence would be intolerable, but so might be a life of unrelenting action and motion.

3 | If sleep were not an agency of change we would wake every day exactly as we were when we went to sleep. But every day is convulsed by a death and a resurrection. Even when there is no trace of the dream that we can put into words, some feeling that the dream has engendered may linger and affect the mood of a whole waking day. That we have dreamed deeply and richly, and know it but cannot express it, tells of the mystery of our unconscious processes of thought. All that went on in the night and is already lost to us in the morning! Our nights are enchanted.

4 | Being awake distinguishes itself from dreaming because it is an experience we can begin to share through art and science – through the mediation of the physical world which is predictable and shareable in certain ways. The laws of physics do not hold in dreams.

5 | 'Composing is like dreaming,' says the composer Harrison Birtwistle. Before a new composition is performed it exists, he says, like a photograph that has been torn into a thousand pieces, shown to him one scrap at a time. Schubert heard little of his orchestral music performed. As what did it exist for him? A performance takes music out of the symbolic world of the score and the mind and into the physical world, where it is transformed.

6 | For centuries philosophers and poets have wondered if life is but a waking dream. Now there seems to be some scientific evidence to support such a claim. It is possible to train yourself both to feel as if you are awake in a dream and to be aware that you are dreaming. In lucid dreaming, bizarrely, the realisation that you are dreaming makes the dream more vivid. The neuroscientist Stephen LaBerge ascribes the sense of vividness to an increase in the intensity of attention on being in the present. He says: 'Right now, if you were to realize the miracle of consciousness and to be *here, now,* you'd have a similar experience.' Such an experience would be like waking up while still being in a dream, except that you would be waking up in waking life – lucid living. He might just as well be describing Buddhist enlightenment. There are probably many more states of consciousness than we generally allow. We tend to think of wakefulness fading into types of unconsciousness, but there are also different states of wakefulness. There are days when we feel we are sleepwalking through our lives, and days when we feel truly alive.

Life is about farting around, and don't let anyone tell you differently.
*Kurt Vonnegut (1922–2007), novelist*

7 | Physics looks for the grand gesture that heralds a universe. To find human beings we must look the other way, towards the small gesture, the small ceremony, the sidelong glance.

The calming effect, Laurence Sterne notes in *Tristram Shandy*, of walking towards a fish pond.

Fanny Price's dilemma in Jane Austen's *Mansfield Park*: Which chain should she wear with the amber cross given to her by her brother William?

The knack of wearing amber and bending over standard roses.
*Ford Madox Ford, Parade's End*

[Doing nothing] means just going along, listening to all the things you can't hear, and not bothering.
*Christopher Robin, in A.A. Milne's The House at Pooh Corner*

> 'Pooh!'
> 'Yes?' said Pooh.
> 'When I'm – when – Pooh!'
> 'Yes, Christopher Robin?'
> 'I'm not going to do Nothing any more.'
> 'Never again?'
> 'Well, not so much. They don't let you.'
> Pooh waited for him to go on, but he was silent again.
>   ...
> 'Pooh, whatever happens, you will understand, won't you?'
> 'Understand what?'
> 'Oh, nothing.' He laughed and jumped to his feet. 'Come on!'
> 'Where?' said Pooh
> 'Anywhere,' said Christopher Robin.
>   *A.A. Milne (1882–1956), The House at Pooh Corner*

8 | Doing nothing and farting around are skills which are lost if not practised. It is hard to let go of the idea that there are more important things to be done. To be truly sympathetic is to know both that everything is of interest and that nothing really matters.

# SECTION 9

# On memory

How wonderful, how very wonderful are the operations of time, and the changes of the human mind! ... If any one faculty of our nature may be called more wonderful than the rest, I do think it is memory. There seems something more speakingly incomprehensible in the powers, the failures, the inequalities of memory, than in any other of our intelligences. The memory is sometimes so retentive, so serviceable, so obedient – at others, so bewildered and so weak – and at others again, so tyrannic, so beyond control! – We are to be sure a miracle every way – but our powers of recollecting and of forgetting, do seem peculiarly past finding out.
*Jane Austen, Mansfield Park\**

But when nothing subsists of an old past, after the death of people, after the destruction of things, alone frailer but more enduring, more immaterial, more persistent, more faithful, smell and taste still remain for a long time, like souls, remembering, waiting, hoping, on the ruin of all the rest, bearing without giving way, on their almost impalpable droplet the immense edifice of memory.
*Marcel Proust, Remembrance of Things Past*

---

\* An uncharacteristically forceful speech made by Fanny Price to Miss Crawford, who 'untouched and inattentive, had nothing to say'.

1 | Memory only exists in organisms that are able to propel themselves using muscles. The sea squirt uses muscles to move itself about the ocean. At some point in its life it finds a rock on which to plant itself, and there it will remain for the rest of its days. As soon as it is situated it consumes – since they are no longer of any use – its own brain and nervous system.

2 | Memories are not hallucinations. Hallucinations feel real. We confuse the hallucination with normal perception, but we never confuse memories and everyday experience. For sufferers of post-traumatic stress syndrome, however, 'memories' of trauma are typically not only misremembered but re-experienced as a hallucination. The syndrome is devastating because the 'memory' feels more real than the original experience.

> Habit weakens all things; but the things which are best at remind-ing us of a person are those which, because they were insignificant, we have forgotten and which have therefore lost none of their power. Which is why the greater part of our memory exists outside us, in a dampish breeze, in the musty air of a bedroom or the smell of autumn's first fires, things through which we can retrieve any part of us that the reasoning mind, having no use for it, disdained, the last vestige of the past, the best of it, the part which, after all our tears seem to have dried, can make us weep again. Outside us? Inside us, more like, but stored away from our mind's eye, in that abeyance of memory which may last for ever.
> Marcel Proust, Remembrance of Things Past

3 | Where do memories live? What space do they inhabit? Are they some kind of skein of knitted neural pathways? Are they stored in the external world like addresses in the iCloud? In bodies and brains, in history, in culture? What do we become if we lose all our memories? Are we no more than an archive of our memories?

4 | When Jean Cocteau revisited his childhood home, no memories stirred until he crouched low and dragged a stick along the wooden fence. And then the memories came flooding back.

5 | Memories can be retrieved from the brain artificially by stimulating it with electricity. Researchers have discovered that smells help call forth emotional memories, but they do not make the memories any more detailed.

6 | The Roman philosopher Cicero (106–43 BC) credited the Greek poet Simonides (c.556–468 BC) with creating the art of memory. Simonides discovered that visual memory works best. The Greek states-man and general Themistocles (c.524–459 BC) memorised the names of 20,000 fellow Athenians. Medieval scholars memorised whole books. Talmudic scholars from Poland even today use visual techniques to memorise thousands of pages of the Babylonian Talmud.

7 | A reductive evolutionary explanation tells us that a good visual memory helps us remember where we put important things.

8 | An experiment by Anthony De Casper and Melanie Spence, psychologists at the University of North Carolina, suggests that babies have memories from birth. A group of pregnant women read *The Cat in the Hat* by Dr Seuss aloud twice a day for the last six weeks of preg-nancy. While their newborns were being breastfed, a story was played to them through an earphone, either *The Cat in the Hat* or some other story they had not heard before. Ten out of twelve babies altered their speed of feeding when the familiar story began. The speed of feeding did not alter when the unfamiliar story was played. More widely, it is supposed that infants do not remember, because the hippocampus, that part of the brain that controls memory, develops after birth.* London

---

* On the other hand, the singer Daniel Bedingfield says he remembers the trauma of being pulled out of the womb.

taxi drivers have a larger hippocampus than other people of the same age. Birds that have more powerful spatial memories than other birds – for example those that store food, like nuthatches and jays – have a larger hippocampus.

9 | When the gears slip and I coast in neutral, paying no attention to the world, to where I am, who am I? My memory brings the world back to me. The world back, and back to me: two miracles. Sometimes when my mind stays blank too long, when for too long a moment I cannot remember the name of my best friend, then I glimpse the fumbling icy maw of the black blank world. When I am engrossed in memories, again the world disappears: no cup, no table, no room, not anything – not even a self, perhaps. Yet here they all are again, returned with a snap of attention, so not lost at all; the living world hiding, ready to jump back into consciousness with a start, consistent with the world as we remembered it when we last checked.

The playwright Samuel Beckett said that memory presents the world in monochrome. My memory hardly presents the world at all. When I think of the past in a generalised way, there's nothing there. I run down empty corridors with a guttering candle looking in at empty rooms. Occasionally here and there I find a broken chair, some ragged curtains flapping at windows that frame what looks like black night, no stars, only the oppressing weight of the annihilating, infinite emptiness of outside.

10 | Memory is feeble if it is to be judged by its ability to reconstitute the past. We live perpetually startled at the disjunction between what is and what has been. Memories make connections not so much across time as outside time. The connections are made not chronologically but out of puns, surreal juxtapositions, random associations, in the way dreams are. Memory seems to work best when we make some synthetic connection between things. Synaesthesia does this naturally. Nothing in the brain is purely auditory, or purely visual, or purely anything. The brain is highly connected. There is interconnection between different

types of sensory input. Synaesthesia is more common than was once thought, and to some degree we are all synaesthetes. Language suggests it: the loud shirt, the sharp cheese. Synaesthesia seems to be common among the creative. Kandinsky, Pollock, Nabokov and Rimbaud were synaesthetes. Richard Feynman saw letters in colours. Scriabin argued with composer friends about what colours the keys were. He said gold/brown for D major, and red/purple for E flat major. V.S. Ramachandran writes of Francesca, for whom touching denim causes feelings of sadness, wax of embarrassment, silk calm, orange-peel shock, sandpaper guilt. There seems to be no common lexicon between synaesthetes; each lexicon is private.

11 | I come across the words 'pencil case', and memories of the pencil cases of childhood come to mind: precise and imprecise. As I close in on them I find that these memories are clustered about by other memories: of the wooden ruler that for some reason has the same smell as the gas that put me under in the dentist's chair. I remember the taste of the blood that leaked from the crater where the abscessed tooth had been; and now the taste of the cloves that did not take away the pain. I remember the feel of cotton wool pressed on my gums. I see the small bottle of BP olive oil in the medicine cupboard, the little pot of thick yellow paste that was meant to salve chilblains, the frost on the windows as I lie in bed ready to jump out into the cold day as if preparing to dive into freezing water, the rush to pile on as many clothes as quickly as possible to stop the shivering from the cold, the struck match to the gas fire, the bar on the electric fire and the smell of burning dust … The dentist with his red hair, bad breath and false smile, my terror, a collective of sadistic teachers, and the generalised feeling of other childhood terrors pressing in, waiting, ghostly; I could sweep down and attend to them if I chose to, but I choose not to and think of something else.

12 | We collectivise our past in memories. Recent research tells us that what we remember from our early childhood is mostly memories of stories told to us by other members of our family. And apparently even

those memories are so lightly held that we can be persuaded to change the details. Families who get together to discuss the details of some past event, variously remembered by the different members, will probably come to some collective agreement: that, say, the dress was red, not yellow; that the month was June, not April, and so on.

*Ah, yes, I remember it well.*

Memories can be fed. If an accident victim is asked: 'Did the collision occur before or after the zebra crossing?' she will remember the zebra crossing and construct a story even though there was no zebra crossing.

13 | When I was a child, I do not remember how old, I made a determined effort to manufacture a memory that I might hold onto and revisit from my future self. As a child I was hardly aware of having a past, but I could see that there was likely to be a future, and where there was a future there would trail a past leading back to me as I was at that moment. I knew that I must get older, but I did not know how. I thought means were required, as if I might fail to age for lack of the necessary skills. How in just a year could I become as changed as the children in the class above? To be four years older seemed to be an insurmountable obstacle. To be an adult was to be laughably and entirely other. I would have to leave myself behind, continually leave selves behind. I wondered if a memory might serve as a rope to throw out to my older self, a rope that I could twitch from time to time from the future and remember my younger self. Myself at both ends: forever the child of that moment at one end, and always changing – for as long as there was rope to unravel – at the other. What memory could hold the weight of me let down into time? I wondered if I should make myself cry, or create some other dramatic gesture, but somehow I knew the memory had to be absolutely true, not falsely dramatic. The memory need be nothing more than the moment exactly as it was: just me sitting on my single bed. I see the bedspread. I remember the feel of it still, the vividness of the orange material, the ridges of the piping, I remember that it was called a candlewick bedspread – actually for years I thought it was

called a camberwell bedspread, until I was corrected recently by a friend, my mistake encouraging me to believe that my memory might be a true one after all. I sat and imagined myself waving to my future self. Are those memories now memories of memories? Sometimes I think I can recall a musty smell, sometimes see the corner of the bedspread where the dog peed, or the ragged wallpaper I had idly scraped from the wall one night as I lay in bed. But these memories may be of other times, a kind of interference signal that has, over the years, attached itself to that original manufactured scene. I think I can still find that child, felt from the inside rather than seen from the outside; the child who was terrified that I might no longer be me, terrified that somewhere along the way the child-me would die. I wave now to that child, yet what has survived most potently is the memory and feeling of that colour orange.

# On faith, belief and truth

The bat that flits at close of eve has left the brain that won't
   believe.
The owl that calls upon the night, speaks the unbeliever's fright.
*William Blake, Auguries of Innocence*

Oh yes, I believe! I believe in what I can see, and what I cannot see.
*Patrick White (1912–90), Riders in the Chariot*

The final belief is to believe in a fiction, which you know to be a
fiction, there being nothing else.
*Wallace Stevens (1879–1955), poet*

1 | The word belief originally meant something closer to the German
*belieben*, to love, a kind of loyalty.

Faith is the substance of things hoped for, the evidence of things
not seen.
*Hebrews 11: 1*

To go from the phantoms of faith to the ghosts of reason is merely
to change cells.
*Fernando Pessoa, The Book of Disquiet*

Faith is a fine invention
When Gentlemen can see –
But Microscopes are prudent
In an emergency.
*Emily Dickinson, poet*


2 | Niels Bohr had a rabbit's foot pinned to his laboratory door.* A visiting scientist expressed surprise that he of all people could possibly believe that a rabbit's foot would bring him luck, to which Bohr replied, 'I was told it would bring me luck whether I believe in it or not.'

> Science, too, has its own brand of faith: the belief that an answer is likely to be true if the results that it predicts are in accord with experimental observation. The more accurate the observation, and the more critical the questions that the experimenter asks, the greater is the scientist's faith in the answer; but this does not mean that the answer is 'right', or that the scientist has discovered 'truth'. Scientific theories can never really be proved true; we simply have faith in them to a greater or lesser extent depending on the difficulty and number of tests that they have passed.
>
> The 'necessary mysteries' that scientists now accept as a result of these tests lie alongside the other set of eternal mysteries that are the province of philosophy and religion. Their reality is overwhelmingly supported by experimental evidence and their existence, to my mind, constitutes very strong evidence for the existence of a world beyond our direct experience.
> *Len Fisher, Weighing the Soul*


3 | For the sake of a set of massive symmetrical particles, for the sake of parallel universes, for the sake of some hidden variable, the particulate nature of the universe might be preserved. What a strange and wonderful belief system material reductionism is.

---

* See page 14.

To say that one needs art or politics that incorporate ambiguity and contradiction is not to say that one then stops recognising and condemning things as evil; however, it might stop one from being utterly convinced of the certainty of one's own solutions. There needs to be a strong understanding of fallibility, and how the very act of certainty, or authoritativeness, can bring disaster.
*William Kentridge, artist and animator*

A quasi-mystical response to nature and the universe is common among scientists and rationalists. It has no connection with supernatural belief.
*Richard Dawkins*

Of all modern delusions, the idea we live in a secular age is the furthest from reality ... Liberal humanism itself is very obviously a religion – a shoddy replica of Christian faith markedly more irrational than the original article, and in recent times more harmful.
*John Gray, philosopher*

And he sigheth deeply in his spirit, and saith, 'Why doth this generation seek after a sign?'
*Mark 8: 11–12*

'I have a belief of my own, and it comforts me.'
'What is that?' said Will, rather jealous of the belief.
'That by desiring what is perfectly good, even when we don't quite know what it is and cannot do what we would, we are part of the divine power against evil – widening the skirts of light and making the struggle with darkness narrower.'
'That is a beautiful mysticism – it is a –'
'Please do not call it by any name,' said Dorothea, putting out her hands entreatingly.

'You will say it is Persian, or something else geographical. It is my life. I have found it out, and cannot part with it …'
*George Eliot, Middlemarch*

4 | What is truth? asked Pontius Pilate, his existential cry propelling him out of time.

Truth, that doubtful onion.
*Patrick White, novelist*

If you're going to tell the truth, make sure you make them laugh; otherwise they'll kill you.
*Neil Simon, playwright*

Truth and belief are uncomfortable words in scholarship. It is possible to define as true only those things that can be proved by certain agreed criteria. In general, science does not believe in truth or, more precisely, science does not believe in belief.

Understanding is understood as the best fit to the data under the current limits (both instrumental and philosophical) of observation. If science fetishized truth it would be religion, which it is not. However, it is clear that under conditions that Thomas Kuhn designated as 'normal science' (as opposed to the intellectual ferment of paradigm shifts) most scholars are involved in supporting what is, in effect, a religion. Their best guesses become fossilized as a status quo, and the status quo becomes an item of faith. So when a scientist tells you that 'the truth is …', it is time to walk away. Better to find a priest.
*Timothy Taylor, archaeologist*

… feeling or truth. Both may be important, but they are not the same thing.
*Richard Dawkins*

Intolerance consists in being so sure of the truth that you want to impose it on everyone else by persuasion, or even by force. We must have an open mind to realise that what suits us doesn't necessarily suit others.
*Matthieu Ricard and Trinh Xuan Thuan, The Quantum and the Lotus*

5 | He is quoting some recent study, the results of which, he says, conclusively proved the inefficacy of prayer. Face distorted by anger, hideous, odd words and phrases, mumbo jumbo, naïvety, Father Christmas, slavery, Church, fear, children. His anger pulls at them like a riptide, ready to take under and away anyone who dares to struggle. Nobody says a word. Shoulders tense and rise, neck sinews twitch involuntarily, they stay as still as they can as if they were small animals that had stumbled by accident upon some voracious predator.

6 | The universe is full of meaning but we do not know what the meaning means, except for when we recognise it as the truth, like music that makes sense even though we cannot say in words what sense.

If 'truth' were whatever I could understand – it would end up being just a small truth, one my size.
*Clarice Lispector, The Passion According to G.H.*

7 | Truth has a gauge finer than the gauge of even our finest nets. Choose your gauge and live accordingly.

# SECTION 11

# On God

## A: The nature of the existence of God

God, you gave us insufficient evidence.
*Bertrand Russell*

The faculty with which we ponder the world has no ability to peer inside itself or our other faculties to see what makes them tick. That makes us the victims of an illusion: that our own psychology comes from some divine force or mysterious essence or almighty principle.
*Steven Pinker*

It does not make much sense to pray to the law of gravity.
*Carl Sagan (1934–96), cosmologist*

1 │ Lucretius believed the world is atoms and motion, but still we must worship, he said, we must worship the atoms and the motion of atoms.

God, he told me, is the progress from chaos to order to human responsibility.
*John Cheever (1912–82), from the short story 'Artemis, the Honest Well Digger'*

I ask myself, if I were God, whether I would have arranged the world in such a way.
*Albert Einstein*

God might have arranged things differently. But he did not.
*James Davidson, The Greeks and Greek Love*

We can act as if there was a God; feel as if we were free; consider nature as if she were full of special designs; lay plans as if we were immortal.
*William James*

2 | Out of belief in God came *The Divine Comedy* and the B Minor Mass. Out of belief in the tooth fairy comes the occasional sixpence. (Paraphrase of an observation made by Rowan Williams, former Archbishop of Canterbury.)

But God may act in subtle ways that are hidden from physical science.
*John Polkinghorne, physicist and theologian*

His everlasting power and deity have been visible ever since the world began, to the eye of reason, in the things that he has made.
*St Paul*

Our world is the first true essay of some infant deity, who afterwards abandoned it, ashamed of his lame performance.
*David Hume, philosopher*

'Miss Potter – where is God?'

'He is everywhere,' replied Miss Potter with dignity.

'But, my dear Maiden,' exclaimed His Highness, planting himself firmly on one of the chairs, 'what good is that to me?'
*J.R. Ackerley, Hindoo Holiday*

Illiterates understand about saints and that saints prove the existence of a benevolent God. The educated are subverted by logic.
*The novelist William Golding (1911–93), in an interview*

3 | The mathematician Pierre-Simon Laplace said of God that he had 'no need of that particular hypothesis'.

4 | It is easy to shoot holes in logical proofs of the existence of God, but most of us long ago gave up believing in that kind of God.

Jewish, Christian and Muslim theologians have insisted for centuries that God does not exist and that there is 'nothing' out there; in making these assertions their aim was not to deny the reality of God but to safeguard God's transcendence.
*Karen Armstrong, writer on comparative religion*

God does not exist. He is being itself beyond essence and existence. Therefore to argue God exists is to deny him.
*Paul Tillich (1886–1965), theologian*

What makes God comprehensible is that he cannot be comprehended.
*Tertullian (c.160–c.225), Christian apologist*

God: the arch-abstraction
*Patrick White*

I don't know whether God exists or not ... Some forms of atheism
are arrogant and ignorant and should be rejected, but agnosticism
– to admit that we don't know and to search – is all right ... When
I look at what I call the gift of life, I feel a gratitude which is in tune
with some religious ideas of God. However, the moment I even
speak of it, I am embarrassed that I may do something wrong to
God in talking about God.

*Karl Popper, in an interview he gave in 1969, on the condition that
it be kept secret until after his death*

5 | We doubters worry that God cannot exist in any meaningful sense
of the word 'exist', and then looking over our shoulders we happen to
catch sight of scientists in the outside lane happily redefining for their
own purposes what existence means.

6 | We cannot resolve the universe. Under scientific examination the
universe becomes more and more abstract, whether examined close to
or looked at from a distance. If not the universe, then why not God?

We do not know what the universe is made out of; or, rather, we
know that it is made out of some kind of nothing. Why, then, would we
expect God to be made out of something?

Science is good at subtle measurement, but these days the public
world has little patience with subtle argument. Science is subtle, why
should God not be subtler? God is not going to appear, no matter how
fine the measurement.

7 | How many angels can fit on a pinhead?* the man in the audience
asked the Australian physicist. More if thin, the physicist replied, fewer
if fat.

---

* Outdoors in full sunlight, a million million photons fall on a pinhead every second.
At night looking up at a faint star, a few hundred photons strike (or do they gently plop
on to?) the retina each second.

8 | In the mid-1980s the Nobel Prize-winning physicist Sheldon Glashow warned that string theory – because it threatened to undermine science by elevating faith – had become 'a new version of medieval theology'.*

9 | We have swapped the mystery of God for the mystery of the laws of nature. Once, we were everywhere with God; but God has fled, leaving behind the Big Bang and a Higgs-type boson.

10 | Materialism reduces God to a 'God of the gaps', hiding in irreducible blood-clotting proteins or in the quantum wave equation, a God allowed to make only occasional outings into the physical world, a peek-a-boo God, a tinkering God restricted to the margins of the world, behind the scenes with an oilcan. Newton worried that the universe might collapse in on itself under gravity. He thought God must intervene to keep the stars apart. God looks on the material world, and sometimes He sighs and sometimes He laughs.

11 | To reduce God to, say, a figure only present at the creation of the universe requires of God that He bow down to the role science offers Him. But it is not only God who is backed into a corner by the scientific method; human beings are too. If humans did not obviously exist, it is likely science would deny the possibility of them. God can be denied more readily. God is a ghost in the machine, but then so are you.

12 | Stephen Hawking famously wrote in *A Brief History of Time* (1988) that if we ever found a complete physical theory 'then we will know the mind of God', which seems as good a way as any of saying that we will never know the mind of God. Perhaps one day we will know the mind of a fly. Perhaps.

---

* String theorists ask esoteric questions like, how many spheres can be packed into a Calabi-Yau space? The answer is 317,206,375. The confirmation of the answer by two separate routes is support for string theory.

13 | The philosopher Jean Baudrillard (1929–2007) once said that the only significant events in the universe are the Big Bang and the Apocalypse. In a physical universe, in which all the significance is weighted to the beginning and the end, humans, who come somewhere in the middle, are skipped over.

14 | Stephen Jay Gould pointed out that if evolution was run twice there might well not be human beings, nor any God to see Himself reflected in humans. On the other hand, nor would there be any scientists to see themselves reflected in their methodology. As Rowan Williams has pointed out, perhaps God needs us as much as we need Him.

15 | There is something provincial about reductive scientific descriptions of the ineffable. They make the universe sound like some sprawling dormitory town, large and rather boring overall but with a few defined areas of interest: perhaps a museum here and a park there. They want us to enjoin in the mystery of the universe, but only on their terms, as if the mystery can only be a certain kind of mystery, whereas true mystery, whatever it is, vanishes unless it is unfettered.

16 | It has been suggested that God might be a being in another universe that compressed a particle of matter to such a high density that a new universe of space and matter was created out of the negative energy of the gravitational field. God is reduced to one of the lads, playing a prank in some other part of the multiverse. Other arguments of this sort include the speculation that God might have left a message encoded in our DNA, like the word BRIGHTON through a stick of rock, or perhaps the message is to be found in the cosmic background radiation, as a sophisticated kind of sky-writing.

17 | If we knew how to make a universe we would have shown that universes can be simulated. But that raises the question: is this a simulated universe that we are living in? If it is, then what we know about

the universe is cast into doubt, including our knowledge that universes can be simulated. This is just the kind of nerdy speculation I enjoy, but I can't help noticing that scientists who claim to despise philosophy and theology love this kind of speculation too.

## B: God of the synapses

1 | Religion has been dismissed as genetic error, neural misfiring or cancerous meme.

2 | The reduction of religious experience to neurology is what William James called medical materialism.

3 | 'Not enough,' the fellow guest sitting next to me at dinner said when I told her about my breakdown and the occasional moments of bliss that it precipitated. 'Jesus came to me once', she said – 'in the kitchen where I was peeling potatoes. I turned around and there he was. I was furious. It's too little too late. Bugger off I said, and he disappeared. Afterwards I told myself he was just an odd firing pattern in my brain.'

But then, where else can anything be except as a firing pattern in the brain? The brain is the junction of all our sensory input, and the instigator of all our responses, and yet still it is not who we are.

> First one then another [voice] presses forward to my shoulder to speak above the engine's voice ... [or they] come out of the air itself, clear yet far away, travelling through distances that can't be measured by the scale of human miles ... conversing and advising me on my flight, discussing problems of my navigation, reassuring me, giving me messages of importance in ordinary life.
> *The American aviator Charles Lindbergh (1902–74), describing his experiences while making the first non-stop trans-Atlantic flight in the 'Spirit of St Louis'*

4 | Sometimes the earth shakes and the ground undulates like a shaken rug. Sometimes when I see the moon rise above the horizon of the sea I feel the earth turn and I lose my balance. Even when time and space become unmoored, when we see visions, and events occur that are impossible, how quickly, afterwards, do the fissures in the space-time continuum close and we find our material footings once more.

Witnesses are only as convincing as their stories. We are well-heeled rationalists most of the time. When what is being described is not rational the story had better be good. Even when we are moved, awed or titillated we remain in doubt, most of us, most of the time. Art can be convincing. Music – surely the most immaterial of all the arts, mute to words, blind to images – seems to me the most convincing evidence (of what of course I cannot quite say), pointing as it does towards meaning, as science does, without the need to trouble purpose.

> What does it matter that it is an abnormal tension, if the result, if the moment of sensation, remembered and analysed in a state of health, turns out to be harmony and beauty brought to their highest point of perfection, and gives a feeling, individual and undreamed of till then, of completeness, proportion, reconciliation, and an ecstatic and prayerful fusion in the highest synthesis of life?
> *Prince Myshkin in Dostoevsky's The Idiot, describing what it is like to experience an epileptic attack*

> A shower of phosphenes in transit across the visual field, their passage being succeeded by a negative scotoma.
> *The neurologist Oliver Sacks accounts for the visions of Hildegard of Bingen (1098–1179), mystic and composer*

> Just as our primary wider-awake consciousness throws open our senses to the touch of things material, so it is logically conceivable that if there be higher spiritual agencies that can directly touch us,

the psychological condition of their doing so might be our possession of a subconscious region which alone should yield access to them.
*William James*

## C: Religion

*Religiare*, Latin: to bind together.

The poet invents the metaphor, and the Christians live it.
*R.S. Thomas (1913–2000), poet*

1 | *Hal* is an Old English word from which hale, health, whole and holy are derived. *Sely* is an early form of the word silly, and that once meant holy. By the 1500s the word had entirely lost its original meaning.

We are in the universe and the universe is in us. I don't know any deeper spiritual feeling than those thoughts.
*Neil de Grasse Tyson, astrophysicist*

That curious sense of the whole residual cosmos as an everlasting presence, intimate or alien, terrible or amusing, lovable or odious, which in some degree every one possesses.
*William James*

I could swallow landscapes, and swill down sunsets, or grapple the whole earth to me with hoops of steel. But the world is so impassive, silent, secret.
*W.N.P. Barbellion*

2 | What we think is out there enables us to negotiate what actually is out there, which is a reality beyond terror and bliss, of which even a glimpse would be enough to destroy us. In the end, maybe it is not the

pressing weight of the material world that threatens to crush us, but the sublime and annihilating power of the immaterial.

If we had a keen vision and feeling of all ordinary human life, it would be like hearing the grass grow and the squirrel's heart beat, and we should die of that roar which lies on the other side of silence. As it is, the quickest of us walk about well wadded with stupidity.

*George Eliot, Middlemarch*

The metaphorical or pantheistic God of the physicists is light years from the interventionist, miracle-wreaking, thought-reading, sin-punishing, prayer-answering God of the Bible.

*Richard Dawkins*

To him, one can do honour in a forest, a field – or merely by gazing up at the ethereal vault, like the ancients. My God is the God of Socrates, of Franklin, of Voltaire, of Béranger! My credo is the credo of Rousseau! ... I have no use for the kind of God who goes walking in his garden with a stick, sends his friends to live in the bellies of whales, gives up the ghost with a groan and then comes back to life three days later! Those things aren't only absurd in themselves, Madame – they're completely opposed to all physical laws. It goes to prove, by the way, that priests have always wallowed in squalid ignorance and have wanted nothing better than to drag the entire world down to their own level.

*Gustave Flaubert, Madame Bovary*

3 | For the philosopher Spinoza (1632–77), God and nature are the same reality. God is not transcendent. The divine is in the world.

My belief is theistic not pantheistic, following Leibniz rather than
Spinoza.
*Kurt Gödel (1906–78), mathematician and philosopher*

4 | Gödel, Heisenberg, Schrödinger and Planck believed in the divine
mind. Einstein spoke of the Old One.

The most beautiful emotion we can experience is the mystical. It
is the sower of all true art and science. He to whom this emotion
is a stranger ... is as good as dead.
*Albert Einstein*

The religious impulse in its truest sense seems to me to be about
that awareness of how extraordinary everything is. Everything.
*Frederick Turner, historian*

5 | The word *olé*, called out in bullfights after some dangerous move
has been elegantly executed, is a corruption of 'Allah'. The shout
acknowledges God, glimpsed in the actions of an artist.

Religion, whatever it is, is man's total reaction upon life, so why
not say any total reaction upon life is a religion?
*William James*

6 | Which in this sense makes Richard Dawkins a religious. I can't help
but see him as a kind of Muscular Christian out of his time. He would
have made a fine cardinal, and but for his outspokenness might have
made Pope.

I told him I wanted to find eternal truths, to help understand why
things are the way they are. Naïve and blustery, for sure ...
*The theoretical physicist Brian Greene, recalling his younger self*

Their religious faculties may be checked in their natural tendency to expand, by beliefs about the world that are inhibitive, the pessimistic and materialistic beliefs, for example, within which so many good souls, who in former times would have freely indulged their religious propensities, find themselves nowadays, as it were, frozen; or the agnostic vetoes upon faith as something weak and shameful, under which so many of us today lie cowering, afraid to use our instincts.

*William James*

# On eternity

1 | Every age heralds the end of days. Our own age has predicted the end of almost everything, from history, poverty, painting and music to sexuality and food. The scientific age, perhaps uniquely among the ages of human history, places itself at the beginning of a process that lasts forever: the ball of growing scientific knowledge rolls forward irreversibly into the future. Scientific progress has no end point in time – which might explain why, at their most extreme, scientists sound like Pangloss and naturalists like Eeyore.

2 | According to the most proselytising of the futurists, some of us who are alive today might live forever. The diseases of old age multiply as our physical bodies deteriorate. If we are to live forever it is of ageing we must cure ourselves, not of disease. There are those who claim that this is not only possible, but that it might be achieved in the near future. If by any chance this book survives so long, are you, dear reader, 850 years old passing for thirty-five? Are you saddened at the prospect of the earth's end some billions of years in the future, regretful that the universe might eventually exhaust itself in, say, $10^{100}$ years' time?

3 | Can we ever outsmart our diseases? The cancer cell, the virus, the bacterium continue to evolve in a changing world. If we are to control them we must hope that human imagination and the scientific method can trump both the random complexities of evolution and the effects of our man-made environment.

4 | Can we repair the earth? Optimistic (realistic?) scientists say yes, so long as we know what kind of machine the earth is, and intervene swiftly and dramatically. Pessimistic (realistic?) naturalists foresee a world exhausted by the unstoppable juggernaut of scientific and economic progress.

5 | How long is long enough? Adam lived 930 years, and might have lived forever but for the Fall. Perhaps for cavemen even their short lives were long enough: that cold, that fear, those parasites. Keats died aged twenty-five, yet at twenty he exhorts us to let time alone, to be more like the flower than the bee. Samuel Beckett said of old age that he had been waiting for it all his life. Eos asked that her lover Tithonus be granted eternal life, but forgot to ask that he be eternally young. Loathsome old age pressed down on him forever. And yet even if he had looked young, that is not the same as being young. Science continually renews itself, but could any human being eternally refresh her spirit? Even if we find ways to ensure that our arteries remain supple, the ego hardens with age. Surely at some point I will have had enough of me; everyone else will have had enough of me long before then. How sad to live too long. Perhaps hell is what lasts forever.

> Archimedes will be remembered when Aeschylus is forgotten, because languages die and mathematical ideas do not. 'Immortality' may be a silly word, but probably a mathematician has the best chance of whatever it may mean.
> G.H. Hardy, A Mathematician's Apology

> A mathematician … has no material to work with but ideas, and so his patterns are likely to last longer, since ideas wear less with time than words.
> Ibid.

6 | Most of us do not remember Archimedes for his theories but for leaping out of his bath and running naked down the streets of Syracuse

crying 'Eureka!' As scientists we care about his theories, *en famille* we do not need to care, and probably do not. In any case, Archimedes' theories would have been discovered by someone else, eventually.

It seems likely that Aeschylus and Archimedes will be remembered for as long as there are humans alive to remember them. But the mathematics that Archimedes uncovered quite possibly exists and has existed forever in that place where the gods and the laws of nature hang out; where human beings are not welcome.

7 | Religious fundamentalists desire eternal life, just not this one. Material fundamentalists want to live forever even if it means living as a computer download.

And in those days shall men seek death, and shall not find it; and shall desire to die, and death shall flee from them.
*The Book of Revelation*

Give us things that are alive and flexible, which won't last too long and become an obstruction and a weariness. Even Michelangelo becomes at last a lump and a burden and a bore.
*D.H. Lawrence, Etruscan Diaries*

For ever! For all eternity! Not for a year or for an age but for ever. Try to imagine the awful meaning of this. You have often seen the sand on the seashore. How fine are its tiny grains! And how many of those tiny little grains go to make up the small handful which a child grasps in its play. Now imagine a mountain of that sand, a million miles high, reaching from the earth to the farthest heavens, and a million miles broad, extending to remotest space, and a million miles in thickness; and imagine such an enormous mass of countless particles of sand multiplied as often as there are leaves in the forest, drops of water in the mighty ocean, feathers on birds, scales on fish, hairs on animals, atoms in the vast expanse of the air: and imagine that at the end of every million years a little bird came to that mountain and carried away in its beak a tiny grain of that sand. How many millions upon millions of centuries would pass before that bird had carried away even a square foot of that mountain, how many eons upon eons of ages before it had carried away all? Yet at the end of that immense stretch of time not even one instant of eternity could be said to have ended. At the end of all those billions and trillions of years eternity would have scarcely begun. And if that mountain rose again after it had been all carried away, and if the bird came again and carried it all away again grain by grain, and if it so rose and sank as many times as there are stars in the sky, atoms in the air, drops of water in the sea, leaves on the trees, feathers upon birds, scales upon fish, hairs upon animals, at the end of all those innumerable risings and sinkings of that immeasurably vast mountain not one single instant of eternity could be said to have ended; even then, at the end of such a period, after that eon of time the mere thought of which makes our very brain reel dizzily, eternity would scarcely have begun.

*James Joyce, A Portrait of the Artist as a Young Man*

The only real voyage of discovery consists not in seeking new landscapes but in having new eyes.

*Marcel Proust*

You know something? We've seen so much, maybe it's better if there are some things that we don't see.

*Peter Matthiessen, The Snow Leopard*

# SECTION 13

# On death

I do not fear death, I had been dead for billions and billions of years before I was born, and had not suffered the slightest inconvenience from it.
*Mark Twain (1835–1910), writer*

Where will you live after death? I will live where the unborn live.
*Seneca (c.4 BC–AD 65), The Trojan Women*

Death, the final boundary between things.
*Horace (65–8 BC), Roman poet*

A rapture of repose.
*Lord Byron (1788–1824), poet*

When you are dead, replied Fernando Pessoa, you know everything, that's one of the advantages.
*José Saramago, The Year of the Death of Ricardo Reis*

1 | For Aristotle the laws of nature describe a universe striving towards perfection. Evolutionists are happier with the idea of perfectibility than of absolute perfection. Evolution may go backwards and forwards, but in the long run it does tend to make things better. Better, not optimal. Darwin believed 'that man in the distant future will be a far more perfect creature than he is now', by which he presumably meant still far

from perfect, but more perfect. Philosophers like Henri Bergson (1859–1941) and Teilhard de Chardin (1881–1955) believed that evolution was leading humanity toward some understanding of its unity with all of creation. This is a noble thought I wish I could share. On those days when I stand 'facing straight into the strong keen wind of understanding' I am all too aware how that sentence of Darwin's ends. Even these more perfect creatures, and indeed all sentient beings, are eventually 'doomed to complete annihilation'. Everything is doomed to die, even the universe. Life is profligate, but it is death that drives evolution. In the far distant future, not only will there no longer be any plants, kangaroos, insects or us; the material of the universe will have spread so far apart that the universe itself will die of the cold.

2 |  In ancient Egypt, death was on the other side of a door. The afterworld was much like this world, but with fantastical elements like walls of iron and trees of turquoise. Barley grew taller there, but not by much. In ancient Greece the afterworld was a dull place of mild torment, of shades wandering about morosely. Discovered in the 1970s in tombs in southern Italy are paintings of handsome naked youths diving into the air, said to represent the soul as it plunges into death.

3 |  She died some 3,300 years ago. She was no more than eighteen years old, a slender-waisted girl about five feet three inches tall, with fair, rather short hair. She was buried on a summer's day. A small yarrow flower was laid on the rim of the coffin before the lid was fixed in place. She wore a short blouse and a belted, knee-length cord skirt. The belt buckle is of bronze and decorated with spirals. A comb made of horn hung on her belt. She wore two bronze arm rings and had a thin ring through her ear. Next to her head is a box made of bark containing a bronze awl and a hairnet. By her feet there is a small bucket also made of bark that once contained a mixture of beer and wine made from wheat and cranberries sweetened with honey. In a cloth bundle by her side are the cremated remains of a child aged about five or six. It seems unlikely, given her age, that the child could have been hers. Perhaps the

child was a sibling who died at the same time. We do not know the young woman's name, but today she is known as the Egtved Girl. The younger child has no name.
(An elaboration of a label in a museum in Copenhagen.)

Whenever I see a dead body, death seems to me a departure. The corpse looks to me like a suit that was left behind. Someone went away and didn't need to take the one and only outfit he'd worn.
*Fernando Pessoa, The Book of Disquiet*

The lonely shepherd's answer to the question what he and his family did without a doctor – 'We just dies a nat'ral death.'
*Eleanor G. Hayden (1865–1964), travel writer*

I do not know if it has ever been noted before that one of the main characteristics of life is discreteness. Unless a film of flesh envelops us, we die. Man exists only insofar as he is separated from his surroundings. The cranium is a space traveller's helmet. Stay inside or you perish. Death is divestment; death is communion. It may be wonderful to mix with the landscape, but to do so is the end of the tender ego.
*Vladimir Nabokov, Pnin*

4 | Consciousness dims from states of wakefulness into states of death. There is cardiac death, cardiopulmonary death, apparent death, somatic death, brain-stem death, whole-brain death, legal death.

If the world is rationally constructed and has meaning, then there must be such a thing.
*Kurt Gödel, on his belief in an afterlife*

Nirvana is nothing.
*Christopher Hitchens (1949–2011), journalist and author*

I think it's a bit like the whole dance of Shiva thing, that you think you're an aloof spectator watching the universe, but actually you're just part of the cosmic ebb and flow of the world. If you think you're part of the ebb and flow of the cosmos, and there's no separate little soul, inspecting the world, that's going to be extinguished – then it's ennobling. You're part of this grand scheme of things.
*V.S. Ramachandran*

I don't think I need to be at one with the universe.
*Richard Gregory (1923–2010), neuropsychologist*

And there, quite suddenly and unexpectedly, I slipped a gear, or something like that. There was not me and the landscape, but a kind of oneness: a connection as though my skin had been blown off. More than that – as though the molecules and atoms I am made of had reunited themselves with the molecules and atoms that the rest of the world is made of. I felt absolutely connected to everything. It was very brief, but it was a total moment.
*Sara Maitland, A Book of Silence*

There was no longer man and boat, but a man-boat, a boat-man.
*Bernard Moitessier (1925–94), yachtsman*

Tantra concerned itself with the totality of existence, the apprehension of the whole universe within man's being ... Tantra might be interpreted as the practice of mankind's earliest religious intuition: that body, mind and nature are all one.
*Peter Matthiessen, The Snow Leopard*

Up the coast a few miles north, in a lava reef under the cliffs, there are a lot of rock pools. You can visit them when the tide is out. Each pool is separate and different, and you can, if you are fanciful, give them names, such as George, Charlotte, Kenny, Mrs. Strunk. Just as George and the others are thought of, for convenience, as individual entities, so you may think of a rock pool as an entity; though, of course, it is not. The waters of its consciousness – so to speak – are swarming with hunted anxieties, grim-jawed greeds, dartingly vivid intuitions, old crusty-shelled rock-gripping obstinacies, deep-down sparkling undiscovered secrets, ominous protean organisms motioning mysteriously, perhaps warningly, toward the surface light. How can such a variety of creatures co-exist at all? Because they have to. The rocks of the pool hold their world together. And, throughout the day of the ebb tide, they know no other.

But that long day ends at last; yields to the night-time of the flood. And, just as the waters of the ocean come flooding, darkening over the pools, so over George and the others in sleep come the waters of that other ocean; that consciousness which is no one in particular but which contains everyone and everything, past, present, and future, and extends unbroken beyond the uttermost stars. We may surely suppose that, in the darkness of the full flood, some of these creatures are lifted from their pools to drift far out over the deep waters. But do they ever bring back, when the daytime of the ebb returns, any kind of catch with them? Can they tell us, in any manner, about their journey? Is there, indeed, anything for them to tell – except that the waters of the ocean are not really other than the waters of the pool?

*Christopher Isherwood (1904–86), A Single Man*

5 | No longer a drop of water but the ocean. The Sufi poet Rumi (1207–73) described life as being like a boat slowly filling up with water. When full, the boat simply falls away and we become one with the limitless ocean. Such descriptions do not make me feel any more at ease with the idea of death, but they do help me prize life. To be at one with the universe is all very well, but it isn't this. Whatever death is and whatever may come afterwards, death is a separation, even if it turns out, paradoxically, to be an end to separation. That it is a paradox is no comfort.

6 | If death is an expansion into everything, then death is nothing at all. If I could escape my ego and put my faith in the world of atoms, I might see that there is no death. Nothing is destroyed in the material world. There are just processes changing energy from one form into other forms. We are ghosts who have failed to account for our existence in the material world. It is hardly surprising that we have also failed to account for our non-existence.

'Who would want to be at one with a stone?' I used to think, dismissively, of Buddhism. Now I sometimes wonder, what could it mean not to be at one with the universe? If separation as a thing were possible even for a moment, the separation must exist forever. The world is of a piece. The glass on the table is only as separate as we wish it to be. Look at the glass closely and we rush back through aeons of time and find ourselves at one with the glass, a patch of energy ready to expand at the Big Bang.

No, there is no comfort here either. I cannot escape my ego. I believe in the world of things. It is where I live.

7 | Why not?, the Buddhist monk said to me in reply to my assertion that whatever came afterwards it could not be this. I think I know what he meant: that I had fallen into a dualist trap of my own devising. If the world is not divided into 'this' and 'that' – if the separation between the rock and the table is at bottom an illusion – how then can I claim to know that the 'that' of death is different from the 'this' of life?

I think of *Slaughterhouse-Five*,* of Billy Pilgrim who slips the egoistic bonds of time and finds past and future laid out eternally existent, the universe as Einstein understood it to be. He escapes the single track of the trammelled ego that runs from birth to death only ever moving forwards. Is this what death is: an ego let loose? A self revisiting the moments of a life, and building them up, eternally, like chords in a symphony? Might Hell be the regret for not having made more or better moments to choose from? Or perhaps Hell is the desire to have

---

* Kurt Vonnegut's novel of 1969.

had different end-stops: if only I had been born a little earlier, died a little later; and Heaven is making do with what there is, what there was. That this. It's a story, anyway, and what choice is there – so long as we live – but to tell the best stories we can?

8 | And yet, although I would like to believe that time is an illusion as physics currently has it, I suspect that time is for the living. Time is real. The universe evolves in time. Everything is as real as everything else, at every scale. Each moment exists and every one of them offers the possibility that something entirely new that has never happened in the universe before might happen *now*. We live in order to experience change, and death is an end to that. No more moments. In death I will become a function of other people's thoughts, a poor substitute indeed for being alive. She will never be forgotten, they said at the funeral. They will never be forgotten, they say of the slain. They lie. Forgetting is what we are good at. Farewell to this life forever. I expect death will be much as it was before birth, and I have forgotten what that was like.

9 | The family gathers around the hospital bed; witnesses. There is a sense of suppressed excitement in the air that no one acknowledges. A man is about to die.

And here it comes, the conjuring trick much anticipated, that moment in time when their father leaves the room without leaving his bed; in his stead a puppet-replica, a piece of perfectly articulated machinery – what was living flesh still miraculous even in its mechanical reduction.

Grief hits the witnesses as hard as a punch from a boxer. Invisible forces beyond the reach of the ego contort them, as if they were trees grown long years on some blasted heath.

The once biological body of their father gives itself over to physics and chemistry. The body sighs its last sighs, a shocking hint of reanimation that is merely gases readjusting, as they must, between un-living barriers.

The time comes, as it must, when the body is taken away; the shape of a father, husband, grandfather, cut out of the world; reduced now to the outline of zipped plastic. The family leaves to live on.

But in the night there is a stirring. The flesh reanimates itself, and what was dispatched comes back. Sometimes, simply, there are flicker-ings – beyond the sensitivities of our best machines, even in the draughty corridors of death, in the ashes of its fire – which catch and flame back into life. The grandfather, father, husband returns, as if he had merely changed his mind.

He tells a story – the usual one – of a tunnel and bright light, of the bliss of extinction. I was headed one way, he says, and wouldn't have turned my head, except that faintly I heard the voices of those I had left behind. Who? Who? Which of us?, his family ask. That's the strangest thing. I heard each of you distinctly, and yet I knew that in some way it was the call of everyone I had ever known, perhaps even the call of everyone and everything there has ever been. I loved the grass and everything equally, he says. I knew you, recognised you, loved each and every one of you.

A shadow of apprehension fell on the family. They felt insulted, generalised. In the years to come, no matter how he tried to explain, they would not, could not understand. He became depressed and his family became first puzzled and then angry, and puzzled at their anger and so angrier. Both sides – and that was how it was, like being on opposite sides – felt let down. His family had suffered the embarrassment of getting what they thought they had wanted; until the day he died again, they never quite forgave him for coming back. Is that why the dead ignore the living, because the worlds are too different?

10 | Oliver Lodge (1851–1940), an English physicist, left a secret message hidden within seven envelopes, one inside the other. He said that in death he would try to tell four famous mediums, instructed to assemble for the purpose, what the secret message said. He died, the Oliver Lodge Posthumous Test Committee convened, the four

mediums were quizzed. After a while the mediums refused to answer any more questions and walked off stage.

11 | On his way to the guillotine, the French chemist and nobleman Antoine Lavoisier (1743–94) told his assistant he would perform one last experiment: he would attempt to blink after his head had been severed. The assistant was instructed to count how many times.

> When I die and go to heaven, there are two matters on which I hope for enlightenment. One is quantum electrodynamics and the other is the turbulent motion of fluids. And about the former I am really optimistic.
> *Attributed to the mathematician Horace Lamb (1849–1934) on his deathbed; also attributed to the physicist Werner Heisenberg (1901–76) on his*

> Why the owl? Why the newt? Why snow on a Douglas fir? Why the river? Why the sound of the river – from far away, from up close? Why the sky for birds to fly through, to fall through? Why the stars so far away? Why loneliness? Why happiness? Why emptiness that won't be filled but so easily fills me? Why the red rust of the garden gate? Why the tree with the hole in it? Why tiny orange lichens?
> *Steve Edwards, Breaking Into the Back Country*

12 | The scientist, as he takes his last breath, smiles to himself. 'I was right, there is nothing. But what a nothing!'

# On humility

My hunch is that most of the time intellectual humility takes one further than arrogance.
*Siri Hustvedt, The Shaking Woman*

We don't know a millionth of one percent about anything.
*Thomas Edison (1847–1931), inventor*

We are learning more and more about less and less.
*Paul Davies, physicist*

He has made all things beautiful in their time, and has put eternity into men's hearts, except that no man will find out the work of the Lord from beginning to end.
*Ecclesiastes*

Moreover, if anyone thinks nothing is to be known, he does not even know whether that can be known, as he says he knows nothing.
*Lucretius*

Such a small speck of creation believing it is capable of comprehending the whole.
*Murray Gell-Mann, physicist*

Every man takes the limits of his own field of vision for the limits
of the world.
*Arthur Schopenhauer*

Not to be absolutely certain is one of the essential things of
rationality.
*Bertrand Russell*

We live amongst riddles and mysteries – the most obvious things,
which come in our way, have dark sides, which the quickest sight
cannot penetrate into; and even the clearest and most exalted
understandings amongst us find ourselves puzzled and at a loss in
almost every cranny of nature's works …
*Laurence Sterne, Tristram Shandy*

1 | The philosopher Montaigne called it *la peste*, the idea that humans
know anything. In retirement in Bordeaux, he wore a pewter medallion
inscribed with the words, '*Que sais-je?*' – What do I know?

2 | We do not know anything. Karl Popper thought this was the most
important philosophical insight there has ever been.

3 | Socrates insisted he had nothing to teach. His students did not go
to learn facts but to learn how to think, how to change their minds.

4 | Without wisdom, writes the literary theorist Stanley Fish, science
would be 'unaided reason and a progress that has no content but, like
the capitalism it reflects and extends, just makes its valueless way into
every nook and cranny'.

5 | That we are not at the centre of anything is science's declaration of
humility.

6 | The Copernican principle is a powerful imprecation to be put alongside other powerful imprecations: Remember you must die; Be still and know I am the Lord.

7 | Science and monotheistic religions believe in progress through time towards some final outcome: a first or second coming, an Apocalypse, the heat death of the universe, a final physical account of the universe written in mathematics.

8 | Materialism is a response to monotheism: both believe that there are fundamental truths that underlie reality; that there are eternal verities, unchanging qualities of the universe, certainties.

9 | Material fundamentalism is allied to religious fundamentalism; both deny metaphor and embrace literalism. Literalism is not a quality limited to overly ardent fans of the Bible or the Koran.

10 | Deep faith questions, otherwise it is merely indoctrination. In an authoritarian world in which beliefs are insisted on, the search for truth becomes the search for certainty. But there is no certainty. Even a belief in uncertainty held too closely becomes certainty of a kind.

11 | At its best monotheism is a series of injunctions and paradoxes. Moses pleaded three times on the day of his death that the Bible be read as poetry. Christ spoke in parables. He bent down and wrote symbols in the sand. He said, Let him who is without sin cast the first stone. Religions are written in metaphor and parable. It is puzzling that Biblical fundamentalism is even possible.

12 | Science asks: What would the world look like without the agency of God? Monotheistic religions ask: If there is a God, what might the world be?*

13 | Both science and religion are capable of acknowledging that the world is always beyond our understanding.

14 | Where religious and scientific extremism meet is in their failure to believe in human beings.

15 | We might fall into despair were we to bind ourselves too tightly to our reductive scientific descriptions. The subtleties that make us human are in perpetual danger of being snuffed out by the weight of the material world.

16 | When science got going in the seventeenth century, meaning – which had once been a quality of our internal world – became a quality of the external world that was to be found out by measurement. We became observers of an external reality where meaning was meant to reside. But as intently as we have looked, and as much as we have found out, we have found no meaning there. It is not surprising then that we find ourselves cut off from the natural world. And yet one of the most powerful qualities of the scientific method – as it gifts us the material world in its wake – is that it continually refines what lies outside itself. Reductive materialism is Cerberus-headed. It looks three ways: deep down towards laws of nature, outwards to unified phenomena that rest on those foundations, and in a third direction: to the beyond, to whatever remains mysterious. If you believe that what remains mysterious is almost everything, then rather than sidestep mystery, reductive materialism actually points it up.

---

* The scientist differs from the religious believer in this: that she may pick up her belief in the scientific method when she arrives at the laboratory, and put it to one side again when she returns home.

17 | The scientific method constantly points to a universe, to human beings, to a God beyond its understanding. What the laws of nature might have to become as science progresses and encompasses more and more phenomena will always be beyond our wildest imaginings. When it comes to it, the difference between mystery and mysticism may not be worth arguing over.

18 | A philosopher once told the poet Anne Carson about the existence of a German phrase that means that which cannot be avoided, *das Unumgängliche*. What cannot be avoided? Light, the universe, God, human beings.

19 | The universe has no bottom. And if there is no getting to the bottom of the universe why would we believe that we will ever get to the bottom of what it is to be a human being?

20 | In science methodology trumps individualism and personality. It is a collective endeavour. Science reminds us to be humble. When it forgets its own imprecation, when science means to rule or to vanquish, the world is imperilled.

21 | The shadow-side of humility is the elevation of ignorance. As Richard Dawkins has pointed out, 'One of the bad effects of religion is that it teaches us that it is a virtue to be satisfied with not understanding.' Yet we might delight in what we know and still know that we know nothing, and delight in that too.

If what we know is as nothing compared with what there is to know, or what we will ever know, does the unseen world not then 'outweigh' the measured world? We might put our faith in the unseen, not as a way of elevating ignorance but as an act of humility. We might peer with delight into our small pool of recently collected knowledge, knowing that behind us the unknown regions stretch away to the horizon. Can we say that after so short a span of investigation we know anything much?

22 | The human perspective may be impossible to remove. Privilege seeps in everywhere. Objectivity runs out and we come face to face with ourselves as subjective content of the world, mysteriously unaccounted for. We may never know truly what it is to be a fly, let alone a proton. (Or do I mean a proton, let alone a fly?) We may never know what it is to be another human being. This is not to say that trying to find out is worthless. Materialism is evidence that the attempt goes somewhere. Art, laughter, shared meals are evidence that empathy is worth pursuing. It is possible to hold these understandings in balance with the knowledge that the world cannot be made out of anything, and that true empathy is unattainable.

23 | Fundamentalism of any kind is death to the human spirit. Only when we know that we human beings matter can we begin to live up to what we are, and what we might be.

Algebra is applied to the clouds; the radiation of the star profits the rose; no thinker would venture to affirm that the perfume of the hawthorn is useless to the constellations. Who, then, can calculate the course of a molecule? How do we know that the creation of worlds is not determined by the fall of grains of sand? Who knows the reciprocal ebb and flow of the infinitely great and the infinitely little, the reverberations of causes in the precipices of being, and the avalanches of creation? The tiniest worm is of importance; the great is little, the little is great; everything is balanced in necessity; alarming vision for the mind. There are marvellous relations between beings and things; in that inexhaustible whole, from the sun to the grub, nothing despises the other; all have need of each other. The light does not bear away terrestrial perfumes into the azure depths, without knowing what it is doing; the night distributes stellar essences to the sleeping flowers. All birds that fly have round their leg the thread of the infinite. Germination is complicated with the bursting forth of a meteor and with the peck of a swallow cracking its egg, and it places on one level the birth of an earthworm and the advent of Socrates. Where the telescope ends, the microscope begins. Which of the two possesses the larger field of vision? Choose. A bit of mould is a pleiad of flowers; a nebula is an ant-hill of stars. The same promiscuousness, and yet more unprecedented, exists between the things of the intelligence and the facts of substance. Elements and principles mingle, combine, wed, multiply with each other, to such a point that the material and the moral world are brought eventually to the same clearness. The phenomenon is perpetually returning upon itself. In the vast cosmic exchanges the universal life goes and comes in unknown quantities, rolling entirely in the invisible mystery of effluvia, employing everything, not losing a single dream, not a single slumber, sowing an animalcule here, crumbling to bits a planet there, oscillating and winding, making of light a force and of thought an element, disseminated and invisible, dissolving all, except that geometrical point, the I; bringing everything back to the soul-atom; expanding everything in God, entangling all activity, from summit to base, in the obscurity of a dizzy mechanism, attaching the flight of an insect to the movement of the earth, subordinating, who knows? Were it only by the identity of the law, the evolution of the comet in the firmament to the whirling of the infusoria in the drop of water. A machine made of mind. Enormous gearing, the prime motor of which is the gnat, and whose final wheel is the zodiac.

*Victor Hugo (1802–85), Les Misérables*

# Select bibliography

I cannot be forever footnoting my conversation as if I were reading an academic paper. It would be a way, I suppose, of trying to ring-fence what is authentically me. But since most of the time I don't know where my ideas come from, and even where my ideas seem more authentic to me they may be things I heard years before, read years before, overheard on a bus and didn't know except perhaps at the time that they had gone in deeply, I am quite happy to accept myself as the agglomeration of everything I have read, heard or attended or not attended to. I don't know where I begin and end, and I'm happy that way. Long may it continue, though I cannot help but suspect that death will draw me up short as it were. Even then, it will be evidence for the everything that I was in living to the everything that is left by my death, that I was that shape after all, and no other.
*Ford Madox Ford, Parade's End*

J.R. Ackerley, *Hindoo Holiday*, 1932 rev 1952, Penguin Modern Classics 2009

Bryan Appleyard, *Aliens: Why They Are Here*, Scribner 2005

Karen Armstrong, *The Case for God: What Religion Really Means*, Bodley Head 2009

Peter Atkins, *On Being: A Scientist's Exploration of the Great Questions of Science*, Oxford 2011

W.H. Auden, *Collected Poems*, ed. Edward Mendelson, Faber 2004

Marcus Aurelius, *Meditations*, trs Robin Hard, Oxford World Classics 2011

Julian Baggini, *Atheism: A Very Short Introduction*, Oxford 2003

W.N.P. Barbellion, *The Journal of a Disappointed Man*, 1919

W.N.P. Barbellion, *A Last Diary*, 1920

Julian Barbour, *The End of Time: The Next Revolution in our Understanding of the Universe*, Oxford 1999

Jonathan Barnes, *Early Greek Philosophy*, Penguin 1987, revised 2001

John D. Barrow, *The Constants of Nature*, Cape 2002

John D. Barrow, *The Book of Universes*, Bodley Head 2011

Giorgio Bassani, *The Garden of the Finzi-Continis*, 1962, trs Jamie McKendrick, Penguin Modern Classics 2007

Sharon Begley, *The Plastic Mind: New Science Reveals Our Extraordinary Potential to Transform Ourselves*, Ballantine 2007

Alex Bellos, *Alex's Adventures in Numberland: Dispatches From the Wonderful World of Mathematics*, Bloomsbury 2010

Erick Beltrain, *The World Explained: A Microhistorical Encyclopaedia*, Roma 2012

John Berger with Jean Mohr, *A Fortunate Man*, Readers Union 1967

Tim Birkhead, *Bird Sense: What it's Like to be a Bird*, Bloomsbury 2012

Susan Blackmore, *Conversations on Consciousness: What the Best Minds Think About the Brain, Free Will, and What it Means to be Human*, Oxford 2006

David Bodanis, *Electric Universe: How Electricity Switched on the Modern World*, Little, Brown 2005

David Bohm, *Wholeness and the Implicate Order*, Routledge 2002

Jill Bolte Taylor, *My Stroke of Insight: A Brain Surgeon's Personal Journey*, Penguin 2006

James Boswell, *The Life of Samuel Johnson*, 1791, Penguin 2008

Brian Boyd, *On the Origin of Stories: Evolution, Cognition, and Fiction*, Harvard 2009

John Brockman ed., *What We Believe But Cannot Prove*, HarperCollins 2006

Derren Brown, *Tricks of the Mind*, Transworld 2007

Sir Thomas Browne, *Religio Medici*, 1643

Sir Thomas Browne, 'Urn Burial', 1658

Mikhail Bulgakov, *The Master and Margarita*, 1967, trs. Diana Burgin and Katherine Tiernan O'Connor, Vintage 1996

Alice Calaprice, *The Quotable Einstein*, Princeton 1996

Rudolf Carnap, 'Intellectual Autobiography', in *The Philosophy of Rudolf Carnap*, ed. P.A. Schilpp, Open Court 1963

Anne Carson, *Nox*, New Directions 2010

Rita Carter, *The Brain Book: An Illustrated Guide to its Structure, Function, and Disorders*, Dorling Kindersley 2009

Jean-Marie Chauvet, Eliotte Brunel Deschamps, and Christian Hillaire, *Chauvet Cave: The Discovery of the World's Oldest Paintings*, new edn Thames and Hudson 2001

John Cheever, *Collected Short Stories*, Vintage 1981

Brian Clegg, *A Brief History of Infinity: The Quest to Think the Unthinkable*, Constable 2003

Frank Close, *Nothing: A Very Short Introduction*, Oxford 2009

J.M. Coetzee and others, *The Lives of Animals*, Princeton 1999

Edward Conze trs., *Buddhist Wisdom: The Diamond Sutra and the Heart Sutra*, Vintage 2001

Jill Cook, *Ice Age Art: Arrival of the Modern Mind*, British Museum 2013

Arthur Cotterell ed., *The Penguin Encyclopedia of Ancient Civilizations*, Viking 1988

Francis Crick, *Of Molecules and Men*, 1967, Prometheus 2004

Francis Crick, *The Astonishing Hypothesis: The Scientific Search for the Soul*, Scribner 1994

Dalai Lama, The, *The Universe in a Single Atom*, Morgan Road Books 2005

Antonio Damasio, *Descartes' Error: Emotion, Reason, and the Human Brain*, Putnam 1994

Antonio Damasio, *Self Comes to Mind: Constructing the Conscious Brain*, Pantheon 2010

Charles Darwin, *The Voyage of the Beagle*, 1839

Charles Darwin, *On the Origin of Species*, 1859

Charles Darwin, *The Descent of Man*, 1871

Darwin's correspondence can be found online at www.darwinproject. ac.uk

Manuel Davenport, *The Mystery of Morality*, 1994

James Davidson, *The Greeks and Greek Love: A Radical Reappraisal of Homosexuality in Ancient Greece*, Weidenfeld 2007

Paul Davies, *Space and Time in the Modern Universe*, Cambridge 1977

Paul Davies, *God and the New Physics*, Oxford 1976

Paul Davies, *Other Worlds: Space, Superspace and the Quantum Universe*, Dent 1980

Paul Davies, *The Eerie Silence: Are We Alone in the Universe?*, Allen Lane 2010

Richard Dawkins, *The Selfish Gene*, Oxford 1976

Richard Dawkins, *The Blind Watchmaker*, Longman 1986

Richard Dawkins, *River Out of Eden*, Weidenfeld and Nicolson 1995

Richard Dawkins, *Climbing Mount Improbable*, Viking 1996

Richard Dawkins, *Unweaving the Rainbow*, Allen Lane 1998

Richard Dawkins, *The God Delusion*, Bantam 2006

Richard Dawkins, *The Greatest Show on Earth: The Evidence for Evolution*, Bantam 2009

Armand Delsemme, *Our Cosmic Origins*, Cambridge 1998

Daniel Dennett, *Consciousness Explained*, Little, Brown 1991

Daniel Dennett, *Darwin's Dangerous Idea*, Simon and Schuster 1995

Daniel Dennett, *Breaking the Spell: Religion as a Natural Phenomenon*, Viking 2006

René Descartes, *Discourse on Method and Other Writings*, trs F.E. Sutcliffe, Penguin 1968

David Deutsch, *The Beginning of Infinity: Explanations That Transform the World*, Allen Lane 2011

Jared Diamond, *Collapse: How Societies Choose to Fail or Survive*, Viking 2005

Mabel Dodge Luhan, *Edge of Taos Desert: An Escape to Reality*, 1937, University of New Mexico Press 1987

Fyodor Dostoevsky, *The Idiot*, trs Richard Pevear and Larissa Volokhonsky, Everyman Library Classic 2002

David Eagleman, *Incognito: The Secret Lives of the Brain*, Canongate 2011

Terry Eagleton, *The Meaning of Life: A Very Short Introduction*, Oxford 2008

Terry Eagleton, *Reason, Faith and Revolution: Reflections on the God Debate*, Yale 2009

John Eccles, *The Human Mystery*, Springer 1979

John Eccles and Karl Popper, *The Self and its Brain*, Springer 1977

Steven Edwards, *Breaking Into the Back Country*, Bison 2010

Niles Eldredge, *The Triumph of Evolution and the Failure of Creationism*, Holt 2000

T.S. Eliot, *Collected Poems*, Faber 1963

Ralph Waldo Emerson, *The Conduct of Life*, 1860 rev. 1876

Mark Epstein, *Going on Being: Buddhism and the Way of Change*, Broadway 2001

Julian Fellowes, scriptwriter, *Gosford Park* (director Robert Altman) 2001

Charles Ferneyhough, *Pieces of Light: The New Science of Memory*, Profile 2012

Richard Feynman, *Surely You're Joking Mr Feynman*, Norton 1985

Richard Feynman, *The Meaning of It All*, Addison-Wesley 1998

Richard Feynman, *The Feynman Lectures on Physics*, commemorative edn Addison Wesley 1989

Victoria Finlay, *Colour: Travels Through the Paintbox*, Hodder and Stoughton 2002

Len Fisher, *Weighing the Soul*, Weidenfeld 2004

Penelope Fitzgerald, *The Gate of Angels*, HarperCollins 1990

Gustave Flaubert, *Madame Bovary*, trs Geoffrey Wall, Penguin Classics 2003

Anthony Flew with Roy Abraham Varghese, *There is a God: How the World's Most Notorious Atheist Changed His Mind*, HarperCollins 2007

Jerry Fodor and Massimo Piatelli-Palmarini, *What Darwin Got Wrong*, Profile 2010

Joshua Foer, *Moonwalking with Einstein: The Art of Remembering Everything*, Allen Lane 2011

Ford Madox Ford, *Parade's End*, tetralogy pub. 1924–1928, Vintage Classics 2012

Michael Frayn, *Constructions: Making Sense of Things*, Wildwood House 1974

Michael Frayn, *The Human Touch: Our Part in the Creation of the Universe*, Faber 2007

John Gardner, *Grendel*, Vintage 1971

Alfred Gell, *Art and Agency*, Oxford 1998

Gevin Giorbran, *Everything Forever: Learning to See Timelessness*, Enchanted Puzzle 2007

Ivan Goncharov, *Oblomov*, trs Stephen Pearl, Bunim and Brown 2006

Jane Goodall, *The Chimpanzees of Gombe: Patterns of Behavior*, Harvard 1986

Jane Goodall, *My Life with the Chimpanzees*, revised edn Simon and Schuster 2007

Stephen Jay Gould, *Wonderful Life*, Norton 1989

Stephen Jay Gould, *Bully for Brontosaurus*, Norton 1991

John Gray, *Straw Dogs: Thoughts on Humans and Other Animals*, Granta 2002

John Gray, *The Immortalization Commission: Science and the Strange Quest to Cheat Death*, Allen Lane 2011

Brian Greene, *The Elegant Universe*, second edn Vintage 2003

Brian Greene, *The Fabric of the Cosmos*, Knopf 2004

Brian Greene, *The Hidden Reality*, Knopf 2011

John Gribbin, *Science: A History*, Allen Lane 2002

Alan Guth, *The Inflationary Universe*, Cape 1997

G.H. Hardy, *A Mathematician's Apology*, Cambridge 1940

Paul Hawken, *Blessed Unrest: How the Largest Movement in the World Came Into Being and Why No One Saw it Coming*, Viking 2007

Stephen Hawking, *A Brief History of Time*, Bantam 1988

Stephen Hawking and Leonard Mlodinow, *The Grand Design*, Bantam 2010

Heraclitus, *Fragments*, trs Brooks Haxton, Viking 2001

Christopher Hitchens, *God is Not Great*, Atlantic 2007

Christopher Hitchens ed., *The Portable Atheist: Essential Readings for the Nonbeliever*, Da Capo 2007

Paul Hoffman, *The Man Who Loved Only Numbers: The Story of Paul Erdös and the Search for Mathematical Truth*, Fourth Estate 1998

Douglas Hofstadter, *I Am a Strange Loop*, Basic 2007

Bert Holldobler and E.O. Wilson, *The Ants*, Harvard 1990

Richard Holmes, *The Age of Wonder: How the Romantic Generation Discovered the Beauty and Terror of Science*, HarperCollins 2008

Michel Houellebecq, *Platform*, trs Frank Wynne, new edn Vintage 2003

Victor Hugo, *Les Misérables*, trs Charles E. Wilbour, Random House 1993

Nicholas Humphrey, *Soul Dust: The Magic of Consciousness*, Quercus 2011

Siri Hustvedt, *The Shaking Woman or a History of My Nerves*, Henry Holt 2009

Christopher Isherwood, *A Single Man*, 1964, Vintage Classics 2010

William James, *Principles of Psychology*, 1890

William James, *The Varieties of Religious Experience*, 1902

Robert Jastrow, *God and the Astronomers*, Norton 1978

Robert Jastrow, *The Enchanted Loom: Mind in the Universe*, Simon and Schuster 1981

Sir James Jeans, *Physics and Philosophy*, Cambridge 1942

Derrick Jensen, *Listening to the Land: Conversations About Nature, Culture, and Eros*, Chelsea Green 2004

Steve Jones, *Almost Like a Whale*, Doubleday 1999

James Joyce, *A Portrait of the Artist as a Young Man*, 1916, Penguin Modern Classics 2000

James Joyce, *Finnegans Wake*, 1939, Penguin Classics 2012

C.G. Jung, *Memories, Dreams, Reflections*, Collins 1963

Michio Kaku, *Physics of the Future: How Physics Will Shape Human Destiny and Our Daily Lives by the Year 2100*, Doubleday 2011

Eric Kandel, *In Search of Memory: The Emergence of a New Science of Mind*, Norton 2006

Robert Kaplan, *The Nothing That is: A Natural History of Zero*, Allen Lane 1999

Stuart Kauffman, *Reinventing the Sacred: A New View of Science, Reason, and Religion*, Basic 2008

John Keats, *The Letters of John Keats 1814–1821*, ed. Hyder Edward Rollins, Harvard 1958

William Kentridge, *Thinking Aloud: Conversations with Angela Breidbach*, David Krut Publishing 2005

George C. Kohn, *Dictionary of Wars*, Facts on File 1986

Lawrence M. Krauss, *A Universe from Nothing: Why There is Something Rather Than Nothing*, Simon and Schuster 2012

Manjit Kumar, *Quantum: Einstein, Bohr and the Great Debate About the Nature of Reality*, Icon 2008

Ray Kurzweil, *The Singularity is Near: When Humans Transcend Biology*, Viking 2005

Ray Kurzweil and Terry Grossman MD, *Transcend: Nine Steps to Living Well Forever*, Rodale 2009

Nick Lane, *Life Ascending: The Ten Great Inventions of Evolution*, Profile 2009

Jaron Lanier, *You Are Not a Gadget: A Manifesto*, Knopf 2010

Ervin Laszlo and others, *The Akashic Experience: Science and the Cosmic Memory Field*, Inner Traditions 2009

Bruno Latour, *Pandora's Hope: Essays on the Reality of Science Studies*, Harvard 1999

Robert B. Laughlin, *A Different Universe: Reinventing Physics from the Bottom Down*, Basic 2005

Peter A. Lawrence, *The Making of a Fly: The Genetics of Animal Design*, Blackwell 1992

Halldór Laxness, *Independent People*, pub in two parts 1934 and 1935, trs J.A. Thompson, Vintage Classics 2008

James Le Fanu, *Why Us?: How Science Rediscovered the Mystery of Ourselves*, HarperCollins 2009

Jonah Lehrer, *Proust Was a Neuroscientist*, Houghton Mifflin 2007

Richard Lewontin, *Biology as Ideology: The Doctrine of DNA*, HarperCollins 1992

Clarice Lispector, *The Passion According to G.H.*, 1964, trs Idra Novey, New Directions 2012

Lucretius, *The Nature of Things*, trs A.E. Stallings, Penguin 2007

Alister E. McGrath, *Surprised by Meaning: Science, Faith and How We Make Sense of Things*, Westminster John Knox Press 2011

Sara Maitland, *A Book of Silence*, Granta 2008

Lynn Margulis and Dorion Sagan, *What is Life?*, new edn University of California Press 2000

Mark Matousek, *Ethical Wisdom: What Makes Us Good*, Doubleday 2011

Peter Matthiessen, *The Snow Leopard*, 1978, Penguin Classics 2008

Mary Midgley, *The Essential Mary Midgley*, ed. David Midgley, Routledge 2005

A.A. Milne, *Winnie the Pooh*, 1926, Egmont 2001

A.A. Milne, *The House at Pooh Corner*, 1928, Egmont 2009

Joni Mitchell, 'A Strange Boy', © 1976 Crazy Crow Music

Michel de Montaigne, *The Complete Essays*, trs M.A. Screech, new edn Penguin Classics 2003

Jan Morris, *Conundrum*, Faber 2002

Guy Murchie, *The Seven Mysteries of Life: An Exploration of Science and Philosophy*, Houghton 1978

Vladimir Nabokov, *Pnin*, 1957, Penguin Modern Classics 2000

Vladimir Nabokov, *Ada or Ardor: A Family Chronicle*, Weidenfeld and Nicolson 1969, Penguin Modern Classics 2011

Sylvia Nasar, *Grand Pursuit: The Story of Economic Genius*, Simon and Schuster 2011

Friedrich Nietzsche, *The Birth of Tragedy*, trs. Shaun Whiteside, Penguin 1993

Friedrich Nietzsche, *Twilight of the Idols*, trs. R.J. Hollingdale, Penguin 1990

David Nokes, *Samuel Johnson: A Life*, Faber 2009

Roger Penrose, *The Road to Reality*, Knopf 2004

Roger Penrose, *Cycles of Time*, Bodley Head 2010

Irene Pepperberg, *The Alex Studies: Cognitive and Communicative Abilities of Grey Parrots*, Harvard 2000

Irene Pepperberg, *Alex and Me: How a Scientist and a Parrot Discovered a Hidden World of Animal Intelligence – and Formed a Deep Bond in the Process*, HarperCollins 2008

Fernando Pessoa, *The Book of Disquiet*, trs Richard Zenith, Carcanet 1991, Penguin Classics 2002

Susan Petrilli and Augusto Ponzio, *Thomas Sebeok and the Signs of Life*, Icon 2001

Adam Phillips, *Darwin's Worms: On Life Stories and Death Stories*, Faber 1999

Adam Phillips, *Going Sane: Maps of Happiness*, Hamish Hamilton 2005

Steven Pinker, *The Language Instinct*, William Morrow 1994

Steven Pinker, *The Blank Slate: The Modern Denial of Human Nature*, Viking 2002

Steven Pinker, *The Stuff of Thought: Language as a Window into Human Nature*, Viking 2007

Steven Pinker, *The Better Angels of Our Nature: Why Violence Has Declined*, Viking 2011

John Polkinghorne, *The Quantum World*, Longman 1984

John Polkinghorne, *Beyond Science: The Wider Human Context*, Cambridge 1996

Karl Popper, *The Logic of Scientific Discovery*, 1934, trs. 1959, rev. edn Routledge 1977

Karl Popper, *The Open Society and Its Enemies*, Routledge 1945

Marcel Proust, *A la recherche du temps perdu* (1913–1922), trs. as *Remembrance of Things Past* by C.K. Scott Moncrieff and Terence Kilmartin, and subsequently as *In Search of Lost Time* in the revision by D.J. Enright, Vintage 2005. I have also used the translation *In Search of Lost Time*, general editor Christopher Prendergast, Allen Lane 2002

V.S. Ramachandran, *The Tell-Tale Brain: Unlocking the Mysteries of Human Nature*, Heinemann 2011

Lisa Randall, *Warped Passages*, HarperCollins 2005

Lisa Randall, *Knocking on Heaven's Door*, HarperCollins 2011

Martin Rees, *Our Final Hour*, Basic 2003

Ed Regis, *What is Life?: Investigating the Nature of Life in the Age of Synthetic Biology*, Farrar Straus 2008

Matthieu Ricard and Trinh Xuan Thuan, *The Quantum and the Lotus*, Three Rivers Press 2001

Matt Ridley, *Genome*, Fourth Estate 1999

Matt Ridley, *Nature via Nurture*, Fourth Estate 2003

Paul Roberts, *Life and Death in Pompeii and Herculaneum*, British Museum 2013

Marilynne Robinson, *Absence of Mind: The Dispelling of Inwardness From the Modern Myth of the Self*, Yale 2010

David Rothenberg, *Survival of the Beautiful: Art, Science and Evolution*, Bloomsbury 2011

Joan Roughgarden, *Evolution and Christian Faith: Reflections of an Evolutionary Biologist*, Island Press 2006

Mark Rowlands, *The Philosopher and the Wolf: Lessons From the Wild on Love, Death and Happiness*, Granta 2008

Susan Roy, *Bomboozled: How the US Government Misled Itself and Its People into Believing They Could Survive a Nuclear Attack*, Pointed Leaf Press 2010

Rumi, *The Essential Rumi*, trs. Coleman Barks with John Moyne, Castle 1997

Oliver Sacks, *Hallucinations*, Picador 2012

José Saramago, *The Year of the Death of Ricardo Reis*, trs Giovanni
    Pontiero, Harcourt 1991
Christopher Scarre and Brian M. Fagan, *Ancient Civilizations*,
    Prentice Hall, second edn 2003
Erwin Schrödinger, *What is Life?*, Cambridge 1944
Gerald L Schroeder, *The Science of God: The Convergence of Scientific
    and Biblical Wisdom*, Free Press 1997
W.G. Sebald, *The Emigrants*, Harvill 1996
W.G. Sebald, *The Rings of Saturn*, Harvill 1998
Charles Seiffe, *Decoding the Universe: How the New Science of
    Information is Explaining Everything in the Cosmos From Our
    Brains to Black Holes*, Viking 2006
W.C. Sellar and R.J. Yeatman, *1066 and All That*, Methuen 1930
William Shawcross ed., *Counting One's Blessings: Selected Letters of
    Queen Elizabeth the Queen Mother*, Macmillan 2012
Rupert Sheldrake, *A New Science of Life*, J.P. Tarcher 1982
Rupert Sheldrake, *The Science Delusion: Freeing the Spirit of Enquiry*,
    Coronet 2012
David Shields, *Reality Hunger: A Manifesto*, Knopf 2010
Daniel J. Siegel, *The Mindful Brain: Reflection and Attunement in the
    Cultivation of Well-Being*, Norton 2007
Peter Singer, *Writings on an Ethical Life*, Fourth Estate 2000
Simon Singh, *Big Bang: The Most Important Scientific Discovery
    of All Time and Why You Need to Know About it*, Fourth Estate
    2004
Lee Smolin, *Time Reborn: From the Crisis in Physics to the Future of
    the Universe*, Houghton Mifflin 2013
Christopher Southgate and others, *God, Humanity and the Cosmos*,
    second edn Continuum 2005
Wallace Stevens, *Collected Poetry and Prose*, Library of America, 1996
Anthony Storr, *Jung*, Routledge 1991
Leonard Susskind, *The Cosmic Landscape: String Theory and the
    Illusion of Intelligent Design*, Little, Brown 2005
Harriet Swain ed., *Big Questions in Science*, Cape 2002

Raymond Tallis, *Aping Mankind: Neuromania, Darwinitis and the Misrepresentation of Humanity*, Acumen 2011

Michael Taussig, *What Color is the Sacred?*, Chicago 2009

John H. Taylor, *Journey Through the Afterlife: The Ancient Egyptian Book of the Dead*, Harvard 2010

Timothy Taylor, *The Artificial Ape: How Technology Changed the Course of Human Evolution*, Macmillan 2010

Nikola Tesla, *My Inventions and Other Writings*, intro. Samantha Hunt, Penguin 2011

Paul Tillich, *Dynamics of Faith*, Harper and Row 1957

Leo Tolstoy, *Anna Karenina*, trs Richard Pevear and Larissa Volokhonsky, Viking 2000

Leo Tolstoy, *The Death of Ivan Ilyich and Other Stories*, trs Richard Pevear and Larissa Volokhonsky, Knopf 2009

Leo Tolstoy, *Resurrection*, trs Anthony Briggs, Penguin 2009

Leo Tolstoy, *War and Peace*, trs Richard Pevear and Larissa Volokhonsky, Vintage 2008

Angus Trumble, *The Finger: A Handbook*, Yale 2010

Jenny Uglow, *The Lunar Men: The Friends Who Made the Future*, Faber 2002

Max Velmans, *The Science of Consciousness*, Routledge 1996

Kurt Vonnegut, *Slaughterhouse Five*, 1969, Vintage Classics 1991

Wendy Wasserstein, *Sloth*, Oxford 2005

Steven Weinberg, *The First Three Minutes: A Modern View of the Origin of the Universe*, Basic 1977

Steven Weinberg, *Dreams of a Final Theory: The Search for the Fundamental Laws of Nature*, Pantheon 1993

Jonathan Weiner, *Time, Love, Memory: A Great Biologist and His Quest for the Origins of Behavior*, Vintage 1999

Hermann Weyl, *The Philosophy of Mathematics and Natural Science*, trs Olaf Helmer, Princeton 1949

Patrick White, *Riders in the Chariot*, 1961, Vintage Classics 1996

Ken Wilbur, *The Marriage of Sense and Soul: Integrating Science and Religion*, Random House 1998

E.O. Wilson, *Sociobiology: The New Synthesis*, Harvard 1975

E.O. Wilson, *Consilience: The Unity of Knowledge*, Knopf 1998

P.G. Wodehouse, *Summer Lightning*, 1929, Arrow 2008

Stephen Wolfram, *A New Kind of Science*, Wolfram Media 2002

Virginia Woolf, *A Room of One's Own*, 1929, Penguin Modern Classics 2002

Virginia Woolf, *To the Lighthouse*, 1927, Collins Classics 2013

Herman Wouk, *The Language God Talks: On Science and Religion*, Little, Brown 2010

Palle Yourgrau, *A World Without Time: The Forgotten Legacy of Gödel and Einstein*, Basic 2005

Edward Zerin, 'Karl Popper on God: The Lost Interview', in the journal *Skeptic* 1998

Slavoj Žižek, *First as Tragedy Then as Farce*, Verso 2009

# Index of names